Leckie

he education publisher
or Scotland

National 5
CHEMISTRY

Practice Workbook

© 2020 Leckie

001/26112020

10 9 8 7 6

HarperCollins Publishers
Macken House, 39/40 Mayor Street Upper, Dublin 1, Ireland, D01 C9W8

ISBN 9780008446789

Published by
Leckie
An imprint of HarperCollins*Publishers*
Westerhill Road, Bishopbriggs, Glasgow, G64 2QT
T: 0844 576 8126   F: 0844 576 8131
leckiescotland@harpercollins.co.uk   www.leckiescotland.co.uk

This material has previously been published in the following titles:
9780008263584 *National 5 Chemistry Practice Question Book* by Barry McBride
9780008281687 *National 5 Chemistry Practice Papers* by Maria D'Arcy and Bob Wilson.

Publisher: Sarah Mitchell
Project Managers: Harley Griffi ths, Lauren Murray and Fiona Watson

Special thanks to
QBS (layout)
Louise Robb (copyedit)
Dylan Hamilton (proofread)
Bob Wilson (answer check)

Printed and bound in the UK using 100%
Renewable Electricity at CPI Group (UK) Ltd

A CIP Catalogue record for this book is available from the British Library.

Acknowledgements
p.34: Natalia Aggiato / Shutterstock; p.37: Jiri Vaclavek / Shutterstock; p.62: petrroudny43 / Shutterstock; p.79: Almog Ziv / Shutterstock; p.110: Shutterstock

MIX
Paper | Supporting responsible forestry
FSC™ C007454

This book is produced from independently certified FSC™ p to ensure responsible forest management.

For more information visit: www.harpercollins.co.uk/green

ebook

To access the ebook version of this Practice Workbook visit
www.collins.co.uk/ebooks
and follow the step-by-step instructions.

About this book 5

SECTION 1   TOPIC QUESTION PRACTICE

CHEMICAL CHANGES AND STRUCTURE

NATURE'S CHEMISTRY

CHEMISTRY IN SOCIETY

SECTION 2   MIXED EXAM QUESTION PRACTICE

ANSWERS Check your answers online:
www.collins.co.uk/pages/Scottish-curriculum-free-resources

About this book

This Practice Workbook has been designed to help you feel confident about your knowledge, and about exams and assessments. It is presented in two parts to provide maximum support in both understanding and exam experience.

The topic practice section contains lots of graded practice in every single topic you will meet on your course. You can use it to consolidate your learning at any point, and to revise and refresh your knowledge in the run-up to exam time. The questions get gradually more challenging to support and extend your knowledge at the same time.

The mixed practice section then gives you the chance to put that knowledge to use in a format and standard that reflects your exams. If you get stuck on a question, you can review the relevant topic section and then come back to try it again.

Good luck!

National 5
CHEMISTRY

Topic Question Practice
Barry McBride

1A Rate of reaction

Hint Rate of reaction can be increased:

- by increasing the temperature
- by increasing the concentration of a reactant
- by increasing surface area/decreasing particle size
- through the use of a catalyst.

Catalysts are substances that speed up chemical reactions but can be recovered chemically unchanged at the end of the reaction.

1 Which of the following would **not** have an effect on the rate of a chemical reaction?

A concentration of reactants

B temperature of the reaction mixture

C size of the beaker

D particle size of the reactants

2 Which of these factors would speed up a chemical reaction?

A decrease in temperature

B decrease in concentration

C decrease in particle size

D decrease in the volume of reactants

3 Different chemical reactions occur at different speeds.

a List the following reactions in order of the rate of the reaction starting with the slowest reaction.

i chips cooking at 200 °C

ii whole potatoes cooking at 200 °C

iii milk turning sour at room temperature

iv milk turning sour in a refrigerator

b Explain the difference in reaction rate between reactions **i** and **ii**.

c Explain the difference in reaction rate between reactions **iii** and **iv**.

4 A student performed the four experiments shown at room temperature (20°C).

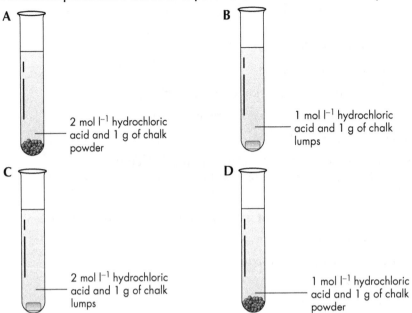

A — 2 mol l⁻¹ hydrochloric acid and 1 g of chalk powder

B — 1 mol l⁻¹ hydrochloric acid and 1 g of chalk lumps

C — 2 mol l⁻¹ hydrochloric acid and 1 g of chalk lumps

D — 1 mol l⁻¹ hydrochloric acid and 1 g of chalk powder

a Which of the four experiments would have the fastest rate of reaction?

b Explain your answer to **a**.

c Which of the four experiments would have the slowest rate of reaction?

d Explain your answer to **c**.

e Explain the effect on the rate of reaction if the experiments were repeated at 40°C.

5 Manganese dioxide catalyses the decomposition (break down) of hydrogen peroxide to form oxygen.

a State the definition of a catalyst.

b If 10 g of manganese dioxide is added to the hydrogen peroxide, what mass of manganese dioxide would remain at the end?

c The manganese dioxide is added in the powdered form rather than a lump.

Explain why this improves the effectiveness of the catalyst.

d The graph below shows the volume of oxygen gas produced over time when the reaction was performed without a catalyst.

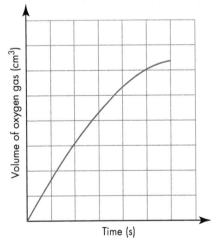

Sketch a new line on the graph to show the results you would expect if the reaction was carried out with manganese dioxide.

6 The volume of gas produced during a chemical reaction can be measured by collecting the gas by displacement of water.

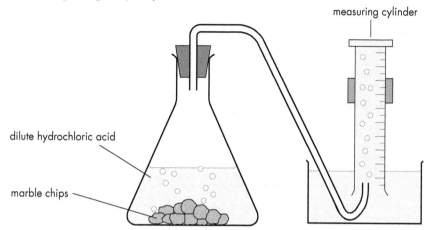

The results below were obtained in the reaction:

Time (s)	0	10	20	30	40	50	60	70	80
Volume of gas (cm³)	0	18	49	63	77	85	89	90	90

a Draw a line graph of the results.

b Mark on the graph the end-point of the reaction.

c The experiment was repeated using the same mass of marble powder.

Sketch a line on your graph to show the expected results.

d Explain why the end-point of the reaction with marble powder is reached before the end-point of the reaction with marble chips.

e Complete the diagram below to show a different method for measuring the gas produced.

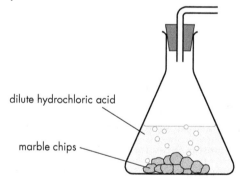

7 The rate of a chemical reaction decreases as the reaction proceeds. Explain why the rate decreases.

8 The progress of a chemical reaction that produces a gas can be monitored by recording the change in mass as the reaction proceeds.

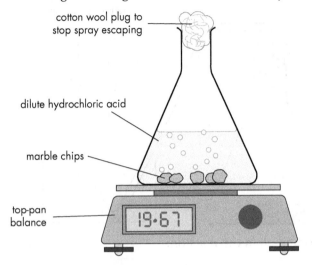

cotton wool plug to stop spray escaping

dilute hydrochloric acid

marble chips

top-pan balance

19·67

The results below were obtained in the reaction.

Time (s)	Mass (g)
10	19·67
20	19·50
30	19·35
40	19·20
50	19·10
60	19·00
70	18·92
80	18·88
90	18·88
100	18·88

a Draw a line graph of the results.

b Mark on the graph the end-point of the reaction.

c Suggest why the mass drops as the reaction proceeds.

d The experiment was repeated using the same mass of marble powder.

Sketch a line on your graph to show the expected results.

e Explain why the end-point of the reaction with marble powder is reached before the end-point of the reaction with marble chips.

1A Rate of reaction

1B Calculating rate

1 Calculate the average rate of reaction, in cm^3 s^{-1}, for the following reactions.

> **Hint**
> $$\text{Average rate} = \frac{\Delta \text{ quantity}}{\Delta \text{ time}}$$

a 40 cm^3 of gas was collected in 20 seconds.

b 15 cm^3 of gas was collected in 60 seconds.

c 30 cm^3 of gas was collected in 30 seconds.

d 10 cm^3 of gas was collected in 40 seconds.

e 20 cm^3 of gas was collected in 35 seconds.

f 8 cm^3 of gas was collected in 32 seconds.

2 Calculate the average rate of reaction for the following reactions. Your answer should include the correct unit.

> **Hint** If the unit is not given in the question then you must provide the correct unit to gain the full marks. This table will help.
>
> | Quantity unit | Time unit | Rate unit |
> |---|---|---|
> | cm^3 | s | cm^3 s^{-1} |
> | g | s | g s^{-1} |
> | l | min | l min^{-1} |

a 1·5 litres of gas was collected in 4 minutes.

b 30 cm^3 of gas was collected in 4 minutes.

c 10 grams were lost in 2·5 minutes.

d 0·4 litres of gas was collected in 120 seconds.

e The mass dropped by 0·8 g in 2 minutes.

f In a period of 10 minutes a total of 1·5 litres of gas was collected.

3 Calculate the average rate of reaction for the following.

> **Hint** The following examples require an extra step: the change in time or change in quantity must be calculated before the rate calculation is performed.

a 20 cm^3 of gas was collected between 20 and 40 seconds.

b The volume of gas collected increased from 10 cm^3 to 50 cm^3 between 15 and 40 seconds.

c The mass decreased from 25 g to 24·1 g between 0 and 60 seconds.

d The volume of carbon dioxide collected increased from 0·1 l to 0·2 l between 2 and 3 minutes.

Your answers should include the correct units.

4

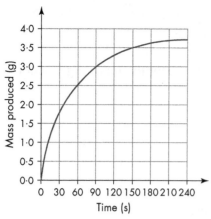

a Calculate the rate, in g s⁻¹, for the first 30 seconds of this reaction.

b Calculate the rate, in g s⁻¹, of this reaction between 30 seconds and 150 seconds.

c Explain why the rate decreased as the reaction progressed.

5 A student carried out an experiment between magnesium metal and excess dilute hydrochloric acid and recorded the results in the table below.

Time (s)	Volume of gas (cm³)
0	0
20	20
40	40
60	50
80	55
100	60
120	60
140	60

a Calculate the average rate of reaction between 20 and 60 seconds.

  Your answer should include the appropriate unit.

b Show by calculation that the average rate was slower between 60 and 100 seconds than it was between 20 and 60 seconds.

c Explain why the rate decreases as the reaction proceeds.

d Draw a line graph of the results and indicate on the graph the end-point of the reaction.

6 A student was investigating the change in mass as calcium carbonate reacted with dilute sulfuric acid and recorded their results in a table.

Time (s)	0	20	40	60	80	100
Mass (g)	80	78	76	75	74	74

a Calculate the average rate of reaction between 0 and 40 seconds.

  Your answer should include the appropriate unit.

b Show by calculation that the average rate was slower between 40 and 80 seconds than it was between 0 and 40 seconds.

c Draw a line graph of the results and indicate on the graph the end-point of the reaction.

7 Many metals react with acid. Hydrogen gas is formed as the metal reacts with the acid to form a salt.

Performing the experiment shown below allows the rate of the reaction to be monitored.

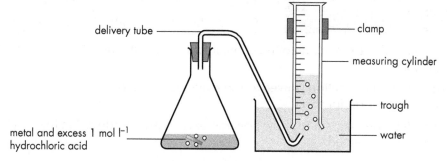

The results below were produced when magnesium metal was used.

Time (s)	Volume of hydrogen gas (cm³)
0	0
20	20
40	42
60	55
80	64
100	68
120	70
140	70

a Draw a line graph of the results.

Use appropriate scales to fill most of the paper.

b Calculate the average rate of reaction between 40 and 80 seconds.

c The rate of the reaction decreases as the reaction progresses.

Suggest a reason for this.

d Use your graph to state the time at which the reaction finished.

8 Hydrogen peroxide (H_2O_2) decomposes to form oxygen gas. A student measured the rate at which hydrogen peroxide decomposes in the presence of a manganese dioxide catalyst and then plotted a graph of the results.

Volume of oxygen gas (cm³)	0	15	25	35	42	50	55	60	62	65	65
Time (s)	0	5	10	15	20	25	30	35	40	45	50

a Calculate the average rate of reaction, in cm³ s⁻¹, for the first 25 seconds of the reaction.

b The rate of the reaction slowed as the reaction proceeded. Suggest a reason for this.

c Complete the diagram to show how the oxygen gas, produced in the reaction, may have been collected.

hydrogen peroxide

1C Atomic structure and the periodic table

1 Complete the statements below by selecting the correct word from the brackets.

   a All known (**elements/compounds/mixtures**) are arranged in the periodic table.

   b Elements in the periodic table are arranged in order of (**increasing/decreasing**) (**mass/atomic**) number.

   c Groups are (**columns/rows/periods**) in the periodic table containing elements with the same number of outer (**protons/electrons/neutrons**).

2 Shown here is part of the periodic table.

1	2											3	4	5	6	7	0
						H											He
Li	Be											B	C	N	O	F	Ne
Na	Mg											Al	Si	P	S	Cl	Ar
K	Ca	Sc	Ti	V	Cr	Mn	Fe	Co	Ni	Cu	Zn	Ga	Ge	As	Se	Br	Kr
Rb	Sr	Y	Zr	Nb	Mo	Tc	Ru	Rh	Pd	Ag	Cd	In	Sn	Sb	Te	I	Xe
Cs	Ba	La	Hf	Ta	W	Re	Os	Ir	Pt	Au	Hg	Tl	Pb	Bi	Po	At	Rn
Fr	Ra	Ac	Rf	Db	Sg	Bh	Hs	Mt	Ds	Rg							

Complete the table below.

Group number	Group name	Description of chemical properties
1		
	Noble gases	
		Very reactive non-metals

3 Isotopes of an element have different:

   A mass numbers   **B** atomic numbers

   C numbers of protons   **D** numbers of electrons.

4 An element has an atomic number of 19 and a mass number of 39.

   How many electrons are there in an atom of this element?

   > **Hint**   In an atom, the number of electrons is equal to the atomic number. This is not true of ions.

   A 19   **B** 20

   C 30   **D** 39

5 An element has an atomic number of 6 and a mass number of 12.

   How many protons are there in an atom of this element?

   A 6   **B** 12

   C 18   **D** 24

6 An element has an atomic number of 11 and a mass number of 23.

How many neutrons are there in an atom of this element?

> **Hint** The atomic number of each element gives us the number of protons that the atoms of that element contain.
>
> The mass number of an element is equal to the atomic number (number of protons) plus the number of neutrons.

A 11   **B** 12

C 23   **D** 34

7 Which of the following electron arrangements is that of an element which has similar chemical properties to potassium?

A 2, 8, 1   **B** 2, 8, 2

C 2, 8, 3   **D** 2, 8, 4

8 Different atoms of the same element have identical:

A numbers of neutrons   **B** atomic numbers

C mass numbers   **D** nuclei.

9 Which line in the table correctly describes a proton?

	Mass	Charge
A	1	−1
B	1	0
C	1	+1
D	0	0

10 The table shows information about some particles.

Particle	Protons	Neutrons	Electrons
A	9	10	10
B	11	12	11
C	15	16	15
D	19	20	18

Which particle is a positive ion?

> **Hint** Ions are charged particles. This means that there is an imbalance in the numbers of electrons and protons. If an ion is negatively charged, there are more electrons than protons, and if it is positively charged, then there are more protons than electrons.

11 Complete the table below.

Particle	Mass	Charge	Location in the atom
Electron			
Proton			
Neutron			

12 Hint The following questions require the use of the SQA data booklet.

a Name each of the elements that are shown by their electron arrangements below:

i 2, 4   **ii** 2, 8   **iii** 2, 8, 8, 2   **iv** 2, 8, 1   **v** 2, 8, 7   **vi** 2, 8, 2

b Which two elements would have similar chemical properties to beryllium?

c Which of the elements is a noble gas?

d Which of the elements is classed as an alkali metal?

13 The nuclide notation of the element sodium is shown.

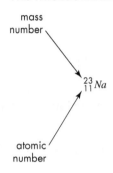

mass number

atomic number

$^{23}_{11}Na$

a State the meaning of the term atomic number.

b State the meaning of the term mass number.

c How many protons, neutrons and electrons does this atom of sodium contain?

d A potassium atom is found to have an atomic number of 19 and a mass number of 39. Write the nuclide notation for this atom of potassium.

e Write the nuclide notation for the following elements.

You may use the SQA data booklet to help you.

i   a lithium atom with 4 neutrons

ii   a silicon atom with a mass number of 28

iii an atom with an atomic number of 12 and 13 neutrons

iv   an atom with 7 protons and 7 neutrons.

14 Give the number of protons, electrons and neutrons contained in each of the following particles:

Hint The atomic number of each element gives us the number of protons that the atoms of that element contain.

The mass number of an element is equal to the atomic number (number of protons) plus the number of neutrons.

a $^{39}_{19}K$ **b** $^{32}_{16}S$ **c** $^{24}_{12}Mg$

d $^{23}_{11}Na^+$ **e** $^{16}_{8}O^{2-}$ **f** $^{40}_{20}Ca^{2+}$

15 **a** Complete the table below. You may use the SQA data booklet to help.

Element	Symbol	Atomic number	Mass number	Protons	Electrons	Neutrons
Calcium	Ca	20				20
Lithium			7	3		
			27	13		
		18	40			
Carbon						6
		16				16
			8	4	4	
	P		31			

b Write the nuclide notation for each of the elements in the table.

16 The nuclide notation can be used to calculate the number of protons, neutrons and electrons that an ion contains. The nuclide notation of a bromide ion is shown.
$^{81}_{35}Br^-$

a Complete the table to show the number of each particle that this ion contains.

Particle	Number
Proton	
Electron	
Neutron	

b A sample of bromine is found to contain two atoms of bromine with different masses: Br-79 and Br-81.

What name is given to atoms of the same element with different masses?

c The relative atomic mass of bromine is 80.

What does this suggest about the relative abundance of these different atoms?

17 **a** Complete the table below. You may use the SQA data booklet to help.

Ion	Symbol	Atomic number	Mass number	Protons	Electrons	Neutrons
Calcium	Ca²⁺	20				20
Chloride			35	17	18	
	Li⁺			3		4
				13	10	14
Fluoride		9	19		10	
	Br⁻	35				46
Sulfide		16	32		18	
	K⁺		40			

b Write the nuclide notation for each of the ions in the table.

c Explain fully why metal elements form positive ions and non-metal elements form negative ions.

18 **a** Explain fully why atoms are electrically neutral.

b Explain fully why ions have an electrical charge.

1C Atomic structure and the periodic table

1D Covalent bonding

1 In a molecule of oxygen the atoms are held together by covalent bonds.

> **Hint** It is important to learn the definition of a covalent bond and to be able to explain how a covalent bond holds atoms together.

a State what is meant by a covalent bond.

b Explain fully how atoms are held together in a covalent bond.

c Which of the following substances contain covalent bonds?

   i H_2O **ii** H_2 **iii** $NaCl$

   iv NH_3 **v** $MgCl_2$ **vi** FeF_2

   vii CaO **viii** PCl_3 **ix** CH_4

2 There are seven elements in the periodic table that are classified as diatomic.

a State what is meant by the term diatomic.

b List the names of the seven diatomic elements.

c Draw a diagram showing all the outer electrons in the molecules of the seven diatomic elements.

> **Hint** In these diagrams always ensure that you have drawn the element symbol at the centre of each atom and that the electrons are clearly drawn on the line of the energy level.
>
> An example for the compound methane (CH_4) is shown.
>
>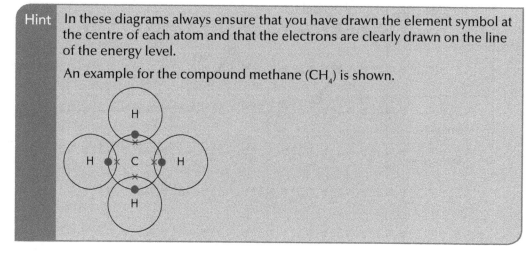

3 Ammonia (NH_3) is a covalent compound.

a Draw a diagram showing the shape of an ammonia molecule.

b Name the shape of an ammonia molecule.

c Draw a diagram showing all the outer electrons in a molecule of ammonia.

4 Listed below are four covalent molecules:

i phosphorus hydride (PH_3)

ii carbon tetrachloride (CCl_4)

iii water (H_2O)

iv hydrogen bromide (HBr)

a Draw a diagram of each molecule showing all the outer electrons.

b Draw and name the shape of each molecule.

> Hint Learn the names of the four shapes of molecules and how to draw them.

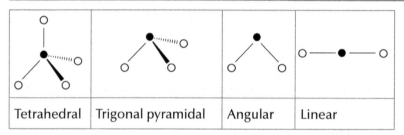

| Tetrahedral | Trigonal pyramidal | Angular | Linear |

5 Complete the table below by inserting the correct name from the word list provided.

ammonia (NH_3), angular, tetrahedral, methane (CH_4), water (H_2O), linear, hydrogen fluoride (HF), trigonal pyramidal

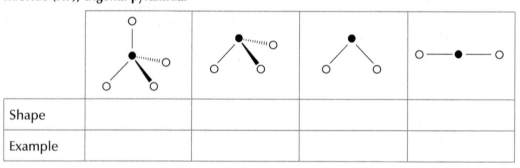

Shape				
Example				

1E Ionic bonding

1 Potassium chloride contains ionic bonds.

> **Hint** It is important to learn the definition of an ionic bond.

  a State what is meant by an ionic bond.

  b Explain fully why atoms form ions.

  c Which of the following substances contain ionic bonds?

    H_2O; H_2; $NaCl$; NH_3; $MgCl_2$, FeF_2; CaO; PCl_3; CH_4

2 Explain fully what is meant by the following terms:

  a ion

  b ionic bond

  c ionic compound.

3 Ions are formed by the loss or gain of electrons.

Write the charge of the following ions:

calcium; fluoride; oxide; aluminium; sodium; bromide; hydrogen; copper(II); phosphide

4 Explain fully why elements in group 0 of the periodic table do not form ions.

5 Complete the following ion-electron equations by adding electrons to each equation:

> **Hint** Use page 10 of the SQA data booklet to help you complete some of these.

  a $Mg \rightarrow Mg^{2+}$

  b $F \rightarrow F^-$

  c $Na \rightarrow Na^+$

  d $Al \rightarrow Al^{3+}$

  e $O \rightarrow O^{2-}$

  f $N \rightarrow N^{3-}$

6 Write ion-electron equations to show the formation of the following ions:

  a Ba^{2+} **b** Li^+

  c K^+ **d** Br^-

  e S^{2-} **f** P^{3-}

7 Write ion-electron equations for the formation of the following atoms:

a Ca   b Na

c Al   d O

e S   f N

g Br   h F

Hint   Remember that the atomic number is equal to the number of protons and that the atomic number of all the elements can be found in the SQA data booklet.

8 Complete the table below. You may wish to use the SQA data booklet to help.

Ion	Number of protons	Number of electrons
Ca^{2+}		
Al^{3+}		
K^+		
Sr^{2+}		
H^+		
I^-		
O^{2-}		
N^{3-}		

1 Which of the diagrams shown represents:

i 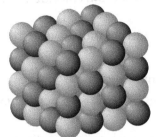 ii  iii

a an ionic lattice

b covalent molecules

c a covalent network?

Hint The table below summarises the properties of substances based on their bonding and structure.

Bonding and structure	Example	Melting point and boiling point	Conduction of electricity
Covalent network	Diamond	Very high	Non-conductors (except for graphite)
Ionic lattice	Sodium chloride (salt)	High	Only when molten or in solution (the ions must be free to move)
Covalent molecule	Carbon dioxide	Low	Non-conductors

2 Titanium tetrachloride ($TiCl_4$) has a melting point of −24 °C.

State the type of bonding in titanium tetrachloride and explain your answer.

3 The properties of three substances, **X**, **Y** and **Z**, were tested and the results recorded in the table below.

	Substance		
Property tested	X	Y	Z
Melting point	High	Very high	Low
Electrical conductivity	Only when molten or in solution	Does not conduct electricity	Does not conduct electricity
Solubility in water	Soluble in water	Insoluble	Insoluble

a State the type of bonding and structure of compounds X, Y and Z.

b Draw a diagram to show how the conductivity of a solution can be tested.

c Explain why substances such as substance X can conduct electricity when molten but not when solid.

4 Sodium chloride (NaCl) is an ionic compound.

   a Explain fully why ionic compounds such as sodium chloride conduct electricity when molten or in solution but not when solid.

   b Explain fully what happens to compounds such as sodium chloride when they dissolve in water.

   c Explain fully why ionic compounds such as sodium chloride have high melting points.

5 Carbon dioxide (CO_2) is a covalent molecular compound.

   a Explain fully why covalent molecular compounds such as carbon dioxide have low melting and boiling points.

   b Explain fully why covalent molecular compounds such as carbon dioxide do not conduct electricity.

6 Silicon carbide (SiC) is a covalent network compound.

   a Explain fully why covalent network substances such as silicon carbide have very high melting and boiling points.

   b Explain fully why covalent network substances such as silicon carbide do not conduct electricity.

1G Chemical formulae

1 Name the elements that each of the compounds listed below contain:

   a sodium chloride   **b** calcium fluoride

   c carbon dioxide   **d** potassium bromide

   e calcium carbonate   **f** magnesium sulfate

   g aluminium nitrate   **h** sodium hydroxide

2 Name the compounds that contain the following elements:

   a lithium and iodine

   b oxygen and aluminium

   c sodium and bromine

   d carbon, copper and oxygen

   e sulfur, oxygen and calcium

   f iron, oxygen and nitrogen

   g phosphorus, potassium and oxygen

   h nitrogen, nickel and oxygen

3 Write the chemical formula of each of the following substances:

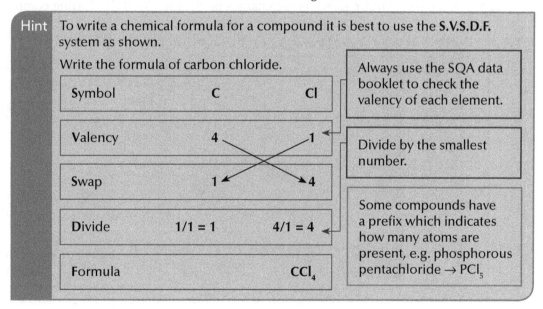

Hint To write a chemical formula for a compound it is best to use the **S.V.S.D.F.** system as shown.

Write the formula of carbon chloride.

Symbol	C	Cl
Valency	4	1
Swap	1	4
Divide	1/1 = 1	4/1 = 4
Formula		CCl_4

Always use the SQA data booklet to check the valency of each element.

Divide by the smallest number.

Some compounds have a prefix which indicates how many atoms are present, e.g. phosphorous pentachloride → PCl_5

   a sodium chloride   **b** calcium fluoride

   c carbon monoxide   **d** potassium bromide

   e carbon dioxide   **f** carbon tetrachloride

   g aluminium oxide   **h** silicon carbide

4 Write the chemical formula of each of the following compounds:

   a copper(II) chloride   **b** zinc(I) fluoride

   c tin(IV) oxide   **d** iron(III) bromide

   e copper(I) oxide   **f** manganese(II) nitride

   g silver(I) sulfide   **h** titanium(IV) chloride

5 Write the chemical formulae of the following compounds:

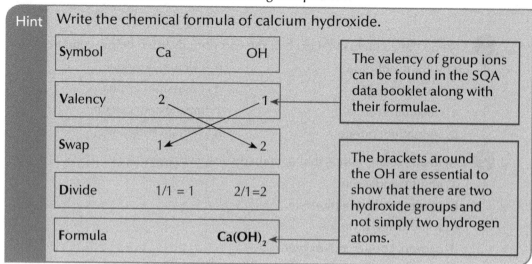

Hint Write the chemical formula of calcium hydroxide.

Symbol	Ca	OH
Valency	2	1
Swap	1	2
Divide	1/1 = 1	2/1=2
Formula		Ca(OH)$_2$

The valency of group ions can be found in the SQA data booklet along with their formulae.

The brackets around the OH are essential to show that there are two hydroxide groups and not simply two hydrogen atoms.

a sodium sulfate

b calcium carbonate

c aluminum phosphate

d lithium hydroxide

e ammonium nitrate

f ammonium phosphate

g ammonium sulfite

h magnesium nitrate

6 Write the formulae showing the charges on the ions for the following compounds:

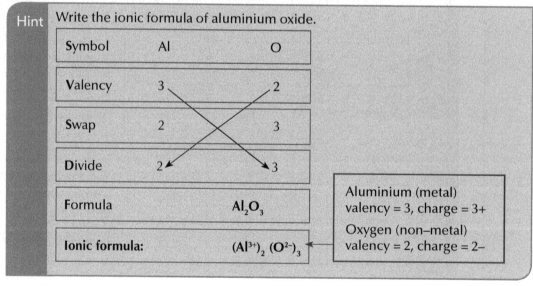

Hint Write the ionic formula of aluminium oxide.

Symbol	Al	O
Valency	3	2
Swap	2	3
Divide	2	3
Formula		Al$_2$O$_3$
Ionic formula:		(Al^{3+})$_2$ (O^{2-})$_3$

Aluminium (metal) valency = 3, charge = 3+

Oxygen (non–metal) valency = 2, charge = 2–

a sodium chloride

b calcium fluoride

c potassium bromide

d magnesium sulfide

e aluminium oxide

f lithium oxide

g beryllium bromide

h aluminium iodide

7 Write the ionic formulae of the following compounds:

a sodium sulfate

b calcium carbonate

c aluminum phosphate

d lithium hydroxide

e ammonium nitrate

f ammonium phosphate

g ammonium sulfite

h magnesium nitrate

1H Balancing equations

1   Balance the following chemical equations:

> **Hint**   **Step 1** – Check that all the formulae are correct using the S.V.S.D.F model as shown in the previous section:
>
>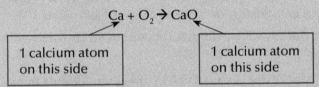
>
> $$Ca + O_2 \rightarrow CaO$$
>
> **Step 2** – Deal with only one element at a time.
>
> $$Ca + O_2 \rightarrow CaO$$
>
> | 1 calcium atom on this side | 1 calcium atom on this side |
>
> **Step 3** – If balancing is required, put the number in front of the substance.
>
> For example, there are two oxygen atoms on the left and only one on the right, therefore multiply the compound on the right by two:
>
> $$Ca + O_2 \rightarrow 2CaO$$
>
> **Step 4** – Check each element again and repeat Step 3 if required.
>
> There are now two calcium atoms on the right and only one on the left, therefore multiply the calcium on the left by two:
>
> $$2Ca + O_2 \rightarrow 2CaO$$

a $Na + O_2 \rightarrow Na_2O$

b $Mg + O_2 \rightarrow MgO$

c $Li + F_2 \rightarrow LiF$

d $Na + HCl \rightarrow NaCl + H_2$

e $Al + Br_2 \rightarrow AlBr_3$

f $CH_4 + O_2 \rightarrow CO_2 + H_2O$

g $K + H_2O \rightarrow KOH + H_2$

h $Fe_2O_3 + CO \rightarrow Fe + CO_2$

i $Ca + HNO_3 \rightarrow Ca(NO_3)_2 + H_2$

j $NaOH + H_2SO_4 \rightarrow Na_2(SO_4) + H_2O$

k $C_2H_6 + O_2 \rightarrow CO_2 + H_2O$

l $C_8H_{18} + O_2 \rightarrow CO_2 + H_2O$

2 Write the balanced formula equation for each of the reactions described below:

a Magnesium metal reacts with hydrochloric acid (HCl) to produce magnesium chloride and hydrogen gas.

b Hydrogen gas is an excellent fuel and burns to produce only water.

c The salt potassium sulfate is produced along with water when potassium hydroxide reacts with sulfuric acid (H_2SO_4).

d Ammonia gas is produced when nitrogen gas reacts with hydrogen gas.

e Lithium hydroxide can be produced by reacting lithium with water. Hydrogen gas is also produced in this reaction.

f Butane (C_4H_{10}) can be used as a fuel for gas barbecues. When it burns, carbon dioxide and water are produced.

g Molten lead(I) bromide can be electrolysed to produce lead metal along with brown fumes of bromine gas.

h Silver metal can be extracted from silver(I) oxide by heating with carbon to produce carbon dioxide along with the silver metal.

i The Hoffmann voltameter can be used to electrolyse hydrochloric acid to produce hydrogen gas and chlorine gas.

 3 Complete the formula equation by writing the formula of molecule **X**.

a $X + 6O_2 \rightarrow 4H_2O + 4CO_2$

b $X + 9{\cdot}5O_2 \rightarrow 7H_2O + 6CO_2$

c $X + 2Br_2 \rightarrow C_4H_6Br_4$

d $X + 2O_2 \rightarrow 2H_2O + CO_2$

e $X + 2HCl \rightarrow CaCl_2 + CO_2 + H_2O$

1I The mole

1 Calculate the **gram formula mass** of each of the compounds shown.

> **Hint** The relative atomic masses can be found in the SQA data booklet.
> Use them and the chemical formula as shown:
>
>
>
> $$(39 \times 2) + 16$$
> $$= 78 + 16 = \textbf{94 g}$$

a Na_2O	**b** MgO	**c** $AlBr_3$	**d** H_2O
e CH_4	**f** Fe_2O_3	**g** C_2H_6	**h** $Ca(NO_3)_2$
i C_8H_{18}	**j** Na_2SO_4	**k** HNO_3	**l** H_2SO_4

2 Calculate the **mass**, in grams, of:

> **Hint** You may wish to use the triangle to help you to remember the relationship between moles, mass and gram formula mass (gfm).
>
> From the triangle:
> mol = mass/gfm; mass = mol × gfm
>
>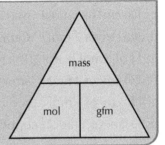

a 2 moles of CO_2

b 0·1 moles of $CaCO_3$

c 0·5 moles of NaCl

d 3 moles of Fe_2O_3

e 0·5 moles of $(NH_4)_2SO_4$

f 10 moles of ethanol (C_2H_5OH)

3 Calculate the **number of moles** present in:

a 16 g of oxygen gas

b 36 g of water

c 49 g of sulfuric acid (H_2SO_4)

d 500 g of $CaCO_3$

e 51 g of NH_3

f 10 g of Li_3PO_4

> **Hint** The concentration triangle will help you remember the connection between concentration, moles and volume. Remember that volume must be in litres.
>
>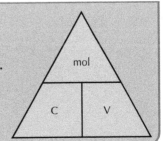

4 Calculate the **concentration**, in mol l^{-1}, of each of the following:

> **Hint** The following calculations require the use of both triangles shown on page 31.

   a 10 g of NaOH dissolved in 100 cm^3 of water

   b 117 g of NaCl dissolved in 500 cm^3 of water

   c 89 g of $AlBr_3$ dissolved in 50 cm^3 of water

   d 52 g of $BaCl_2$ dissolved in 1 l of water

   e 254 g of iron(II) chloride dissolved in 1·5 l of water

   f 20 g of ammonium nitrate dissolved in 100 cm^3 of water

   g 11·1 g of $CaCl_2$ dissolved in 200 cm^3 of water

   h 1 g of KCl dissolved in 50 cm^3 of water

5 Calculate the **mass**, in grams, required to produce each of the solutions below:

   a 50 cm^3 of 2 mol l^{-1} LiOH solution

   b 250 cm^3 of 1 mol l^{-1} NaOH solution

   c 100 cm^3 of 3 mol l^{-1} Na_2CO_3 solution

   d 500 cm^3 of 2 mol l^{-1} $CaBr_2$ solution

   e 1 l of 3 mol l^{-1} iron(III) chloride solution

   f 200 cm^3 of 2 mol l^{-1} ammonium sulfate solution

   g 100 cm^3 of 1 mol l^{-1} sodium sulfate solution

   h 50 cm^3 of 5 mol l^{-1} copper(II) sulfate solution

To answer questions 6–9, the steps shown must be followed.

Calculate the mass of water produced on combustion of 20 g of hydrogen gas.

$$2H_2 + O_2 \rightarrow 2H_2O$$

Step 2 – Calculate the number of moles of H_2 reacted. Remember that hydrogen gas is diatomic so the gfm is 2 and not 1.

2 moles of $H_2 \rightarrow$ 2 moles of H_2O

$2 \rightarrow 2$

$1 \rightarrow 1$

moles = mass/gfm $10 \rightarrow 10$

= 20/2

= 10 moles

mass = moles × gfm

Step 4 – Calculate the mass of H_2O produced using:

= 10 × 18

= 180 g

Step 1 – Establish the molar ratio using the balanced equation.

Step 3 – Use the molar ratio to establish the moles of H_2O produced.

The ratio is one to one; since the number of moles of H_2 reacted is 10 moles, 10 moles of water were produced.

6. Use the balanced equation to calculate the mass of carbon dioxide produced on combustion of 22 g of propane gas.

$$C_3H_8 + 5O_2 \rightarrow 3CO_2 + 4H_2O$$

7. Calculate the mass of copper produced when 25·5 g of copper sulfate reacts with excess magnesium metal.

$$Mg + CuSO_4 \rightarrow MgSO_4 + Cu$$

8. Calculate the mass of hydrogen required to react with nitrogen to produce 85 g of ammonia gas.

$$3H_2 + N_2 \rightarrow 2NH_3$$

9. Calculate the mass of lithium that would react with 8 g of oxygen gas.

$$2Li + O_2 \rightarrow Li_2O$$

1J Percentage composition

> **Hint**
>
> $$\text{Percentage composition} = \frac{\text{Mass of element} \times 100}{\text{Formula mass}}$$

1 Calculate the percentage by mass of calcium in calcium carbonate ($CaCO_3$).

2 Iron can be extracted from iron(III) oxide in a blast furnace.

Calculate the percentage by mass of iron in iron(III) oxide.

3 Ammonium nitrate is a very good fertiliser.

Calculate the percentage by mass of nitrogen in ammonium nitrate.

4 Aluminium can be extracted from aluminium oxide by electrolysis.

Calculate the percentage by mass of aluminium in aluminium oxide.

5 The fertiliser ammonium phosphate (($NH_4)_3PO_4$) contains two essential elements for healthy plant growth. Calculate the percentage by mass of **both** nitrogen and phosphorous contained within ammonium phosphate.

6 Lead(II) fluoride can be used as a catalyst.

Calculate the percentage by mass of lead in lead(II) fluoride.

7 Urea ($CO(NH_2)_2$) is a compound found in urine.

Calculate the percentage by mass of nitrogen in urea.

8 Complete the table to show the percentage by mass of metal in each of the compounds.

Compound	Percentage of metal element
Iron(II) oxide (FeO)	
Potassium sulfate (K_2SO_4)	
Sodium chloride (NaCl)	
Aluminium oxide (Al_2O_3)	
Calcium nitrate ($Ca(NO_3)_2$)	
Lithium fluoride (LiF)	
Strontium chloride ($SrCl_2$)	

9 Which of the following nitrogen-containing compounds has the highest percentage by mass of nitrogen?

a ammonium nitrate

b nitric acid

c ammonium phosphate

d potassium nitride

1K Acids and bases

1 Complete the statements below by selecting the correct word from the brackets.

  a The pH scale is a measure of the (**hydrogen/hydroxide**) ion concentration and runs from below (**0/14**) to above (**0/14**).

  b All acidic solutions have a higher concentration of (**H⁺/OH⁻**) than (**H⁺/OH⁻**) ions and have a pH (**above/below**) 7.

  c All alkaline solutions have a higher concentration of (**H⁺/OH⁻**) than (**H⁺/OH⁻**) ions and have a pH (**above/below**) 7.

  d Dilution of an acidic solution with water will (**increase/decrease**) the (**H⁺/OH⁻**) ion concentration and (**increase/decrease**) the pH towards 7.

  e Dilution of an alkaline solution with water will (**increase/decrease**) the (**H⁺/OH⁻**) ion concentration and (**increase/decrease**) the pH towards 7.

2 Water is pH neutral.

  a Explain fully why water has a neutral pH.

  b Explain what is meant by the term 'dissociates'.

  c Write the equation for the dissociation of water molecules.

3 Which of the following compounds can be classed as bases?

  > **Hint** A base is a substance that neutralises an acid.

  Na_2O; SO_2; $Mg(OH)_2$; $CaCO_3$; CO_2; NH_3; CaO; HCl

4 Soluble metal oxides dissolve in water to produce alkaline solutions.

  Write the balanced chemical equations for the following metal oxides when dissolved in water.

  a sodium oxide

  b lithium oxide

  c barium oxide

  d calcium oxide

  e potassium oxide

5 Complete the two tables below by completing the formulae and ionic formulae of common lab acids and alkalis.

  > **Hint** It is important to learn the formulae of the common lab acids.

Acid name	Formula	Ionic formula
Hydrochloric acid		
Sulfuric acid		
Nitric acid		

Alkali name	Formula	Ionic formula
Sodium hydroxide		
Potassium hydroxide		
Lithium hydroxide		

6 Adding copper carbonate to sulfuric acid can produce copper sulfate crystals like the one shown below.

a Write the formula equation for this reaction.

b State what happens to the pH of the acid solution as the copper carbonate is added to it.

c Describe how to identify the gas produced.

d When the reaction is complete the excess copper carbonate must be removed.
   Draw a diagram of how the excess copper carbonate could be removed in the lab.

e The final step involves removing the water to leave the copper sulfate crystals.
   Draw a diagram of how the water could be removed from the solution in the lab.

7 Complete the following neutralisation equations. There is no need to balance the equations.

Hint	base + acid → salt + water

a sodium oxide + sulfuric acid → _____

b magnesium hydroxide + _____ → magnesium chloride + water

c _____ + nitric acid → calcium nitrate + water + carbon dioxide

d $Li_2O + HNO_3 →$ _____

e $Al_2O_3 +$ _____ $→ AlCl_3 + H_2O$

f $NH_3 + H_2SO_4 →$ _____

8 A student tested the conductivity of two solutions of hydrochloric acid and recorded the results in a table.

Property tested	0·1 mol l⁻¹ HCl	1 mol l⁻¹ HCl
pH	1	0
Conductivity	low	high

a Explain fully why the conductivity of the 1 mol l⁻¹ HCl solution is higher than that of the 0·1 mol l⁻¹ HCl solution.

b Which solution would require the greatest amount of alkali to neutralise?

c Draw a diagram to show how the conductivity of each solution could be tested in the lab.

9 The equation for the reaction of sodium hydroxide and hydrochloric acid is shown.

$$NaOH(aq) + HCl(aq) \rightarrow NaCl(aq) + H_2O(l)$$

a Write the ionic equation for this reaction.

b Name the two spectator ions in this reaction.

c Rewrite the ionic equation omitting the spectator ions.

10 The equation for the reaction of lithium hydroxide and sulfuric acid is shown.

$$H_2SO_4(aq) + 2LiOH(aq) \rightarrow Li_2SO_4(aq) + 2H_2O(l)$$

a Write the equation showing the ionic charges for each ion in this reaction.

b Name the two spectator ions in this reaction.

c Rewrite the ionic equation omitting the spectator ions.

11 The equation for the reaction of potassium oxide and nitric acid is shown.

$$K_2O(s) + 2HNO_3(aq) \rightarrow 2KNO_3(aq) + H_2O(l)$$

a Write the ionic equation for this reaction.

b Name the spectator ion in this reaction.

c Rewrite the ionic equation omitting the spectator ions.

12 The equation for the reaction of calcium carbonate and hydrochloric acid is shown.

$$CaCO_3(s) + 2HCl(aq) \rightarrow CaCl_2(aq) + H_2O(l) + CO_2(g)$$

a Write the ionic equation for this reaction.

b Name the spectator ion in this reaction.

c Rewrite the ionic equation omitting the spectator ions.

1L Titrations

1 Calculate the average volume, in cm³, that should be used from the following tables of titration results.

> **Hint** Only the concordant results (results within 0·2 cm³ of each other) should be used to calculate the average.

a

Titration	Volume added (cm³)
1	9·9
2	10·3
3	10·2

b

Titration	Volume added (cm³)
1	12·2
2	13·0
3	13·0
4	13·1

c

Titration	Volume added (cm³)
1	20·9
2	20·1
3	20·3
4	20·6

d

Titration	Volume added (cm³)
1	18·8
2	19·1
3	19·3

e

Titration	Volume added (cm³)
1	18·0
2	18·8
3	19·5
4	18·8

f

Titration	Volume added (cm³)
1	15·0
2	15·6
3	15·7
4	15·8

To answer the following questions, follow the steps shown.

The balanced equation required for the calculation will be given in the question.

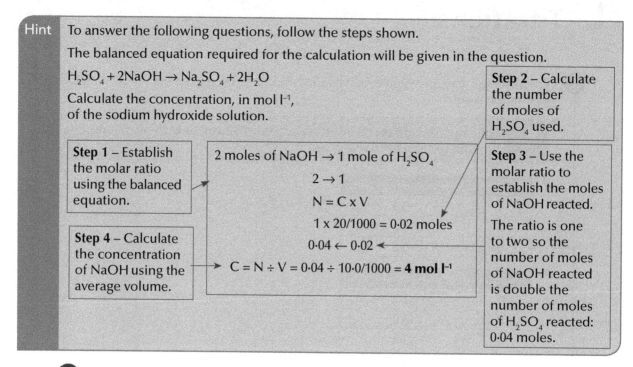

$H_2SO_4 + 2NaOH \rightarrow Na_2SO_4 + 2H_2O$

Calculate the concentration, in mol l⁻¹, of the sodium hydroxide solution.

Step 1 – Establish the molar ratio using the balanced equation.

Step 4 – Calculate the concentration of NaOH using the average volume.

2 moles of NaOH → 1 mole of H_2SO_4

$2 \rightarrow 1$

$N = C \times V$

$1 \times 20/1000 = 0.02$ moles

$0.04 \leftarrow 0.02$

$C = N \div V = 0.04 \div 10.0/1000 = $ **4 mol l⁻¹**

Step 2 – Calculate the number of moles of H_2SO_4 used.

Step 3 – Use the molar ratio to establish the moles of NaOH reacted.

The ratio is one to two so the number of moles of NaOH reacted is double the number of moles of H_2SO_4 reacted: 0.04 moles.

2 A student aims to establish the concentration of a hydrochloric acid solution by titrating it with a standard solution of 1.0 mol l⁻¹ sodium hydroxide.

$$NaOH(aq) + HCl(aq) \rightarrow NaCl(aq) + H_2O(l)$$

a State what is meant by the term standard solution.

b This titration requires the use of an indicator. Explain why an indicator is required.

c The student repeats the experiment until concordant results are achieved.

Explain what is meant by the term concordant results.

d It was established 20 cm³ of the 1.0 mol l⁻¹ solution was required to neutralise 10 cm³ of the hydrochloric acid solution.

Calculate the concentration, in mol l⁻¹, of the hydrochloric acid solution.

3 Vinegar is a dilute solution of ethanoic acid (CH_3COOH). The concentration of ethanoic acid in a famous brand of vinegar can be established by titrating 20 cm³ of the acid against a standard solution of sodium hydroxide 0.1 mol l⁻¹.

$$CH_3COOH(aq) + NaOH(aq) \rightarrow CH_3COONa(aq) + H_2O(l)$$

The results were recorded in the table below.

Titration	Volume added (cm³)
1	9.8
2	10.1
3	10.1
4	10.2

a State what is meant by the term standard solution.

b The titration was performed using a burette, pipette and a standard flask.

Draw a diagram of these pieces of apparatus.

c Calculate the average volume of sodium hydroxide that reacted with the ethanoic acid.

d Calculate the concentration of ethanoic acid in the vinegar.

Your answer should include the appropriate unit.

e Explain fully how an uncontaminated dry sample of the salt sodium ethanoate could be obtained.

4 Calculate the volume of 2 mol l^{-1} potassium hydroxide required to neutralise the following acids:

a 20 cm^3 of 1 mol l^{-1} hydrochloric acid

$$KOH(aq) + HCl(aq) \rightarrow KCl(aq) + H_2O(l)$$

b 10 cm^3 of 1 mol l^{-1} sulfuric acid

$$2KOH(aq) + H_2SO_4(aq) \rightarrow K_2SO_4(aq) + 2H_2O(l)$$

c 10 cm^3 of 1 mol l^{-1} phosphoric acid

$$3KOH(aq) + H_3PO_4(aq) \rightarrow K_3PO_4(aq) + 3H_2O(l)$$

d 12 cm^3 of 1·5 mol l^{-1} ethanoic acid

$$KOH(aq) + CH_3COOH(aq) \rightarrow CH_3COOK(aq) + H_2O(l)$$

5 Calculate the volume of sodium hydroxide required to neutralise each acid:

a 1·0 mol l^{-1} NaOH and 50 cm^3 of 1 mol l^{-1} hydrochloric acid

$$NaOH(aq) + HCl(aq) \rightarrow NaCl(aq) + H_2O(l)$$

b 2·0 mol l^{-1} NaOH and 10 cm^3 of 1 mol l^{-1} sulfuric acid

$$2NaOH(aq) + H_2SO_4(aq) \rightarrow Na_2SO_4(aq) + 2H_2O(l)$$

c 0·5 mol l^{-1} NaOH and 10 cm^3 of 2 mol l^{-1} phosphoric acid

$$3NaOH(aq) + H_3PO_4(aq) \rightarrow Na_3PO_4(aq) + 3H_2O(l)$$

d 1·5 mol l^{-1} NaOH and 15 cm^3 of 2 mol l^{-1} nitric acid

$$KOH(aq) + HNO_3(aq) \rightarrow KNO_3(aq) + H_2O(l)$$

6 Ammonia is a base that reacts with nitric acid to produce the salt ammonium nitrate.

$$NH_3(aq) + HNO_3(aq) \rightarrow NH_4NO_3(aq) + H_2O(l)$$

A titration is performed to establish the concentration of ammonia solution by performing a titration with a 1·5 mol l^{-1} standard solution of nitric acid and the results are recorded in the table below.

Titration	Volume added (cm³)
1	9·8
2	10·1
3	10·1
4	10·2

a Calculate the average volume of nitric acid, in cm^3, that should be used to calculate the concentration of the ammonia solution.

b 20 cm^3 of the ammonia solution was transferred to the conical flask used in the titration.

State the name of the piece of equipment that should be used to accurately measure the 20 cm^3 of ammonia solution.

c Calculate the concentration of the ammonia solution.

Your answer must include the appropriate unit.

d The ammonium nitrate produced is an important fertiliser.

Explain fully how an uncontaminated dry sample of the salt could be obtained.

2A Homologous series

1 The alkanes are a homologous series of saturated hydrocarbons.

> **Hint** It is important to learn the exact definition of the term homologous series.

  a State what is meant by the term homologous series.
  b State what is meant by the term hydrocarbon.
  c Name the first member of the alkanes' homologous series.
  d State what is meant by the term saturated when used in reference to alkanes.
  e Write the general formula for the alkanes.
  f Give a use for alkanes.

2 Copy and complete the table below.

> **Hint** When drawing full structural formulae ensure that all bonds are drawn and that all the hydrogen atoms are shown. Always check that each carbon atom has formed four bonds.

Alkane	Molecular formula	Full structural formula
Methane	CH_4	
	C_4H_{10}	
Hexane		
		H—C—C—C—C—C—C—C—C—H (with H atoms above and below each of the 8 carbons)

3 Using the general formula, complete the formulae of the following alkanes:

> **Hint** Use the general formula or C_nH_{2n+2} to complete the formula, e.g. if an alkane has 10 carbons then it will have $(2 \times 10) + 2 = 22$ hydrogens, giving $C_{10}H_{22}$

  a C_5H_x **b** C_xH_{10} **c** $C_{11}H_x$
  d C_xH_{16} **e** $C_{15}H_x$ **f** C_xH_{30}

4 The alkanes show a pattern in how the physical properties change as the molecular size increases.

> **Hint** Question 4 is a common exam question and you should ensure that you answer it fully.

  a State what happens to the melting and boiling points of the alkanes as the molecular size increases.
  b Explain fully why the melting and boiling points change in this way.

5 The alkenes are a homologous series of unsaturated hydrocarbons.

   a Name the first member of the alkenes' homologous series.

   b State what is meant by the term unsaturated when used in reference to alkenes.

   c Write the general formula for the alkenes.

   d Give a use for alkenes.

6 Copy and complete the table below.

> **Hint** Always make sure that all the carbon atoms have formed four bonds.

Alkene	Molecular formula	Full structural formula
Ethene	C_2H_4	
	C_6H_{12}	
Pent-1-ene		

7 Using the general formula, complete the formulae of the following alkenes:

   a C_3H_x

   b C_xH_{10}

   c $C_{10}H_x$

   d C_xH_{16}

   e $C_{12}H_x$

   f C_xH_{30}

> **Hint** Question 8 is a common exam question and you should be able to answer it fully. The boiling point of the alkanes increases but the flammability decreases as the carbon chain increases. This is because as the size of the carbon chain increases, more intermolecular bonds can form, resulting in higher boiling points and melting points but a lower flammability.
>
> short hydrocarbon chains   long hydrocarbon chains
>
>
> intermolecular bonds   intermolecular bonds

8 The alkenes show a pattern in how the physical properties change as the molecular size increases.

   a State what happens to the melting and boiling points of the alkenes as the molecular size increases.

   b Explain fully why the melting and boiling points change in this way.

9 The cycloalkanes are a homologous series of saturated hydrocarbons.

   a Name the first member of the cycloalkanes' homologous series.

   b State what is meant by the term saturated when used in reference to cycloalkanes.

   c Write the general formula for the cycloalkanes.

   d Give a use for cycloalkanes.

10 Copy and complete the table below.

Cycloalkane	Molecular formula	Full structural formula
Cyclopropane	C_3H_6	
	C_5H_{10}	
Cyclohexane		

11 Using the general formula, complete the formulae of the following cycloalkanes:

   a C_4H_x **b** C_xH_{12} **c** C_8H_x

   d C_xH_{20} **e** $C_{18}H_x$ **f** C_xH_{30}

12 Name the following hydrocarbons:

> **Hint** Naming hydrocarbons is easier if you draw the full structural formula when you are given a shortened formula.

   a

   b

   c

   d

   e $CH_3CH_2CH_2CH_2CH_3$ **f** $CH_3CHCHCH_3$

   g CH_2CHCH_3 **h** $CH_3CH_2CH_2CH_2CH_2CH_2CH_3$

13 The cycloalkanes show a pattern in how the physical properties change as the molecular size increases.

  a State what happens to the melting and boiling points of the cycloalkanes as the molecular size increases.

  b Explain fully why the melting and boiling points change in this way.

14 The alkynes are a homologous series.

  Ethyne H—C≡C—H

  Propyne H—C≡C—C—H (with H above and H below the third C)

  But-2-yne H—C—C≡C—C—H (with H above and below the first C, and H above and below the fourth C)

  a Draw the full structural formula of the fourth member of the alkynes' homologous series.

  b Write the general formula of the alkynes' homologous series.

  c State whether the first or fourth member of the alkynes' homologous series would have the higher melting point.

  d Explain fully your answer to **c**.

2B Addition reactions

Hint There are several types of addition reaction – hydration (addition of water), hydrogenation (addition of hydrogen), halogenation (addition of a halogen). Make sure that you know what all these reactions are.

1 Pent-1-ene is an unsaturated hydrocarbon.

a Draw the full structural formula of pent-1-ene.

b Describe the test used to identify unsaturated hydrocarbons and give the result.

c Name the type of reaction that takes place in the test described in **b**.

2 Hydrogen reacts with ethene to produce ethane in an addition reaction.

a Write the word equation for this reaction.

b Give another name for this addition reaction.

c Draw the full structural formula of the product of this reaction.

3 Markovnikov's rule states that if a compound such as H_2O or HBr is added to an unsaturated compound, the hydrogen atom of the H_2O or HBr will add to the carbon atom which is part of the double bond and already has the most hydrogen atoms attached.

a What name is given to the addition reactions that involve the addition of:

   i HBr   **ii** H_2O

b Taking into account Markovnikov's rule, draw the full structural formulae of the compounds produced when but-1-ene reacts with hydrogen bromide and water.

$$H-\overset{\overset{\displaystyle H}{|}}{\underset{\underset{\displaystyle H}{|}}{C}}-\overset{\overset{\displaystyle H}{|}}{\underset{\underset{\displaystyle H}{|}}{C}}-\overset{\overset{\displaystyle H}{|}}{\underset{\underset{\displaystyle H}{|}}{C}}=\overset{\overset{\displaystyle H}{}}{C} + HBr \rightarrow$$

$$H-\overset{\overset{\displaystyle H}{|}}{\underset{\underset{\displaystyle H}{|}}{C}}-\overset{\overset{\displaystyle H}{|}}{\underset{\underset{\displaystyle H}{|}}{C}}-\overset{\overset{\displaystyle H}{|}}{\underset{\underset{\displaystyle H}{|}}{C}}=\overset{\overset{\displaystyle H}{}}{C} + H_2O \rightarrow$$

c Explain why Markvnikov's rule does not need to be considered in addition reactions involving but-2-ene.

4 Listed below are several examples of addition reactions.

Complete each equation and give the type of each addition reaction taking place.

	Type of addition reaction
$C_2H_4 + H_2 \rightarrow$	
$C_3H_6 + I_2 \rightarrow$	Iodination/Halogenation
$C_5H_6 + H_2O \rightarrow$	
$C_2H_4 + HBr \rightarrow$	Halogenation

5 The alcohol ethanol (C_2H_5OH) can be produced by the hydration of ethene.

   a Write the formula equation for this reaction.

   b Draw the full structural formula of the product.

   c Give another name for this type of reaction.

6 Dehydration is the removing of water from a molecule such as ethanol to form an alkene.

> **Hint** When describing the colour change associated with the bromine test, always ensure that you clearly state that it is the bromine that is changing colour and not the compound tested.

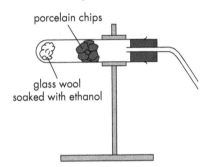

This can be performed in the lab as shown.

porcelain chips

glass wool
soaked with ethanol

   a Suggest the name of the product of this reaction.

   b Complete the diagram to show how the product gas could be collected.

   c The gas collected is tested with bromine solution. State the colour change that would be observed if the gas produced is unsaturated.

7 **a** Name the type of addition reaction that coverts ethene into:

    i an alkane

    ii a dihaloalkane

    iii an alcohol

   b Write a molecular formula equation for each type of reaction.

1 State the name of each of the following branched alkanes.

> **Hint** Follow these steps to name branched alkanes.
>
> **a** Find the longest carbon chain and name as normal, e.g. five carbon atoms means that it is pentane.
>
> **b** Identify the branch and name. (A branch with one carbon is methyl, a branch with two carbons is ethyl, etc.)
>
> **c** Number the carbons in the chain so that the branch is on the lowest possible number.

a

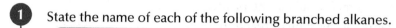

b

c

d

e

f

2 State the name of each of the following branched alkenes:

a
```
    H      H  H
    |      |  |
H—C—C = C—C—H
    |      |
    H      H
    |
 H—C—H
    |
    H
```

b
```
                    H
                    |
                H—C—H
                    |
                H—C—H
                    |
  H  H  H        |  H  H  H  H
  |  |  |        |  |  |  |  |
H—C—C—C = C—C—C—C—H
  |  |           |  |  |  |
  H  H        H  H  H  H
```

c
```
            H
            |
        H—C—H
 H  H      |
 |  |      |
H—C—C—C = C—H
 |  |  |
 H  H  H—C—H
          |
          H
```

d
```
        H
        |
    H—C—H
        |
        H
        |
H—C—C = C—H
    |       |
    H       H
```

e
```
            H
            |
        H—C—H
 H  H      |   H
 |  |      |   |
H—C—C—C = C—C—H
 |  |  |      |
 H  H  H      H
```

f
```
    H            H
    |            |
H—C—H    H—C—H
    |       |    |
    H       H    H
    |       |    |
H—C—C = C—C—C—C—H
    |       |  |  |
    H       H  H  H  H
```

3 Draw the full structural formula **and** give the name of the following hydrocarbons:

a $CH_3CH_2CH_2CH(CH_3)CH_2CH_3$ **b** $CH_3CH(CH_3)CH_3$

c $CH_3CH_2C(CH_3)_2CH_3$ **d** $CH_3CH(CH_3)CH(CH_3)CH_2CH_2CH_3$

e $CH_2CHCH(CH_3)CH_3$ **f** $CH_3CH_2C(CH_3)_2CHCH_2$

4 Draw the full structural formula of each of the following compounds:

a 2-methylbutane   **b** 3-methylhexane

c 2,3-dimethylpentane   **d** 3,3-dimethylheptane

e but-2-ene   **f** 2-methylpent-2-ene

g 3-methylbut-1-ene

2D Isomers

Hint When drawing isomers always ensure that the two structures have the same number of atoms but are arranged differently.

1 Shown below are three isomers of pentane.

Isomer	pentane	2-methylbutane	2,2-dimethylpropane
Structure	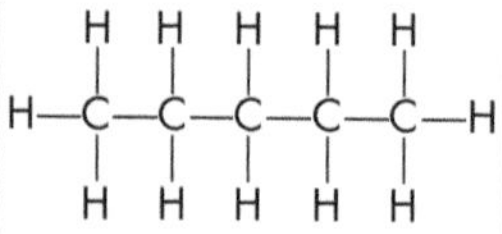	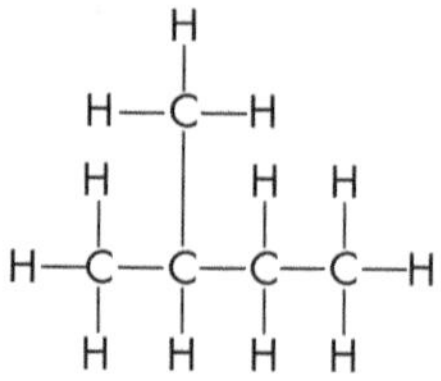	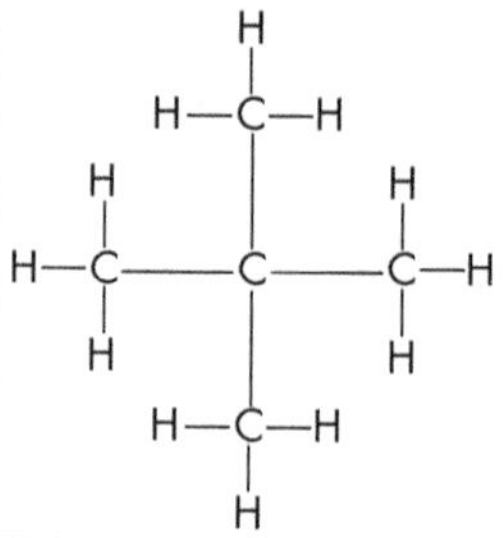
Boiling point (°C)	36	27	11

a State the definition of the term isomer.

b Make a statement linking the number of branches on a molecule and its boiling point.

2 Butane has the structure shown.

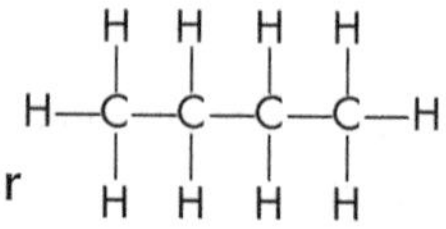

a Draw an isomer of butane.

b Give the systematic name for the isomer you have drawn.

3 1,2-dichloroethane has the structure shown.

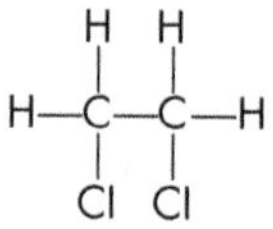

a Draw an isomer of 1,2-dichloroethane.

b Suggest a name for the isomer you have drawn.

4 Draw and name an isomer of each of the following molecules:

a

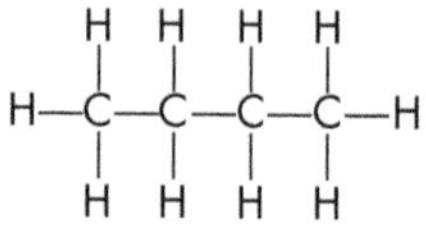

b

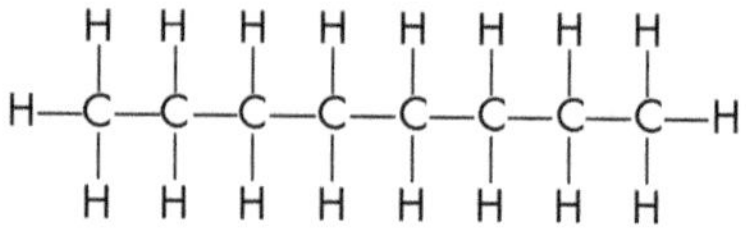

c

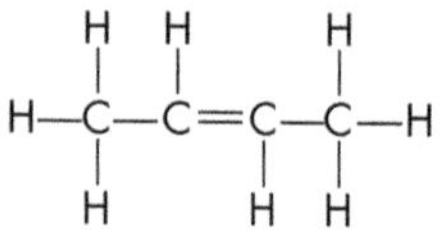

d

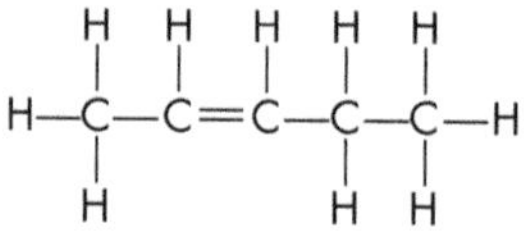

e

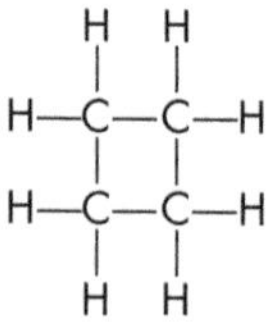

f

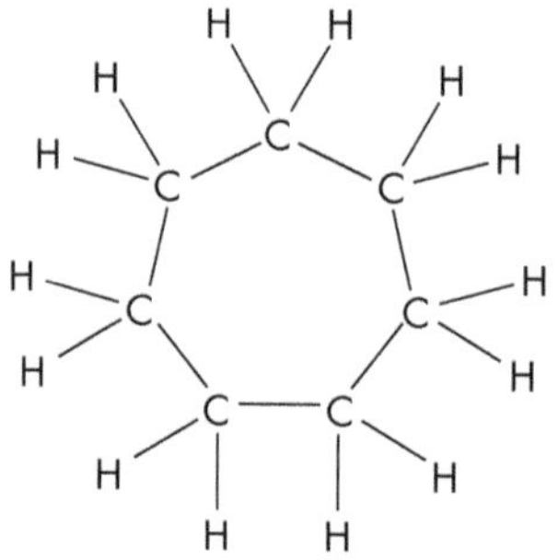

5 Draw and name two possible isomers of molecules with the molecular formulae shown.

a C_5H_{12}

b C_6H_{12}

c C_4H_8

d C_7H_{14}

e C_8H_{18}

6 Which of the following are isomers of hexane?

a 2-methylpentane

b 3-methylhexane

c 3-methylpentane

d 2,2-dimethylpentane

e 2,2-dimethylbutane

f 2,3-dimethylbutane

g cyclohexane

7 Ethanal (CH_3CHO) and 1,2-ethanediol are isomers.

a State what is meant by the term isomer.

b 1,2-ethandiol has the following structure:

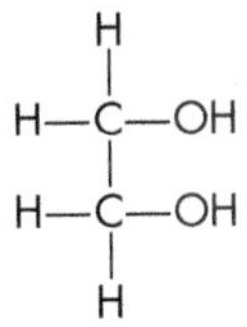

Draw a possible full structural formula of ethanal.

2E Alcohols

1 The alcohols are a homologous series.

> **Hint** Make a list of all the homologous series and their functional groups as this will help you to learn them.

  a State what is meant by the term homologous series.

  b Name the functional group found in alcohols.

  c Name the first member of the alcohols' homologous series.

  d Write the general formula for the alcohols.

  e State two uses of alcohols.

2 Copy and complete the table below.

> **Hint** When drawing the full structural formulae of alcohols, always ensure that the bond from the carbon goes to the O of the hydroxyl group and not the H.

Alcohol	Molecular formula	Full structural formula
Ethanol	C_2H_5OH	
Butan-1-ol		
	$C_6H_{13}OH$	

3 Use the general formula of the alcohols to complete the molecular formulae of the following alcohols:

  a C_3H_xOH

  b C_xH_9OH

  c C_5H_xOH

  d $C_xH_{17}OH$

  e $C_{10}H_xOH$

  f $C_xH_{29}OH$

4 The alcohols show a pattern in how their physical properties change as the molecular size increases.

    a State what happens to the melting and boiling points of the alcohols as the molecular size increases.

    b Explain fully why the melting and boiling points change in this way.

5 Alcohols can be produced industrially by reacting an alkene with water in the presence of a phosphoric acid catalyst.

> **Hint** This is another addition reaction.

$$\underset{H}{\overset{H}{\diagdown}}C = C\underset{H}{\overset{H}{\diagup}} + H_2O \rightarrow$$

    a Copy and complete the equation.

    b Give two names for the type of reaction taking place.

    c Suggest why the use of a catalyst makes the process more economical.

    d Propene can also be used to produce alcohols in the same process. When propene is used two isomeric alcohols are produced. Draw and name the two isomers produced.

6 Shown are the shortened structural formulae of some alcohols.

Draw the full structural formulae and name the alcohols shown.

    a $CH_3CH_2CH_2OH$

    b $HOCH_2CH_2CH_2CH_3$

    c $CH_3CH(OH)CH_2CH_3$

    d $CH_3CH_2CH_2CH(OH)CH_2CH_3$

    e $CH_3CH_2CH(OH)CH_2CH_3$

    f $CH_3CH(OH)CH_3$

7 Draw the full structural formula of each of the following alcohols:

    a ethanol

    b propan-1-ol

    c pentan-2-ol

    d butan-2-ol

    e hexan-1-ol

    f propan-2-ol

    g pentan-1-ol

8 Alcohols can be classified as primary, secondary or tertiary.

In primary alcohols, the functional group is attached to a carbon that has two hydrogen atoms attached to it.

```
    H  H  H
    |  |  |
H—C—C—C—H
    |  |  |
    H  H  OH
```

In secondary alcohols, the functional group is attached to a carbon atom that has one hydrogen atom attached to it.

```
    H  H  H
    |  |  |
H—C—C—C—H
    |  |  |
    H  OH H
```

In tertiary alcohols, the functional group is attached to a carbon atom which has no hydrogen atoms attached to it.

```
        H
        |
   H—C—H
    H  |  H
    |  |  |
H—C—C—C—H
    |  |  |
    H  OH H
```

a Name the functional group contained in all alcohols.

b Give the systematic name for each of the three types of alcohol shown.

c Draw the full structural formula of a secondary alcohol containing four carbon atoms.

2F Carboxylic acids

1 The carboxylic acids are a homologous series.

> **Hint** Carboxylic acids are a family of compounds with the carboxyl functional group (-COOH), for example, propanoic acid.
>
>
> $CH_3\ CH_2\ COOH$

  a State what is meant by the term homologous series.

  b Name the functional group found in carboxylic acids.

  c Name the first member of the carboxylic acids' homologous series.

  d Write the general formula for the carboxylic acids.

  e Give two uses of carboxylic acids.

2 Copy and complete the table below.

> **Hint** Remember that the carbon atom involved in the functional group is still included in the name of the carboxylic acid.

Carboxylic acid	Molecular formula	Full structural formula
Methanoic acid	HCOOH	
Butanoic acid		
	$C_7H_{15}COOH$	

3 Use the general formula of the carboxylic acids to complete the molecular formulae of the following:

  a C_3H_xCOOH **b** C_xH_9COOH

  c C_6H_xCOOH **d** $C_xH_{17}COOH$

  e $C_{10}H_xCOOH$ **f** $C_xH_{29}COOH$

4 The carboxylic acids show a pattern in how their physical properties change as the molecular size increases.

  a State what happens to the melting and boiling points of the alcohols as the molecular size increases.

  b Explain fully why the melting and boiling points change in this way.

5 Vinegar is a dilute solution of ethanoic acid. The concentration of ethanoic acid in vinegar can be established by titration with sodium hydroxide.

$$CH_3COOH(aq) + NaOH(aq) \rightarrow CH_3COONa(aq) + H_2O(l)$$

A titration was performed using a 1·5 mol l^{-1} standard solution of sodium hydroxide and 20 cm^3 of vinegar to establish the concentration of ethanoic acid in the solution of vinegar. The results were recorded in the table shown.

Titration	Volume added (cm³)
Rough	11·0
1	10·6
2	10·7
3	10·8

a Calculate the average volume of sodium hydroxide, in cm^3, that should be used to calculate the concentration of the ethanoic acid solution.

b Calculate the concentration of the ethanoic acid solution.

Your answer must include the appropriate unit.

6 Carboxylic acids take part in neutralisation reactions with bases.

> **Hint**   acid + base → salt + water

a State the change in pH as the carboxylic acid is neutralised.

b State what is meant by the term base when used in this context.

c Complete the neutralisation equations below.

  i   ethanoic acid   + sodium hydroxide   → _____

  ii   butanoic acid   + calcium oxide   → _____

  iii   _____ + lithium oxide   → lithium ethanoate + water

  iv   _____ + sodium hydroxide   → sodium hexanoate + water

  v   propanoic acid + calcium carbonate   → _____

  vi   butanoic acid + sodium carbonate   → _____

  vii $CH_3CH_2COOH + KOH$ → _____

  viii $CH_3CH_2CH_2COOH + Na_2O$ → _____

  ix $CH_3COOH + Li_2CO_3$ → _____

  x   _____ + _____   → $COONa + H_2O + CO_2$

2G Energy from fuels

1. Complete the statements below by selecting the correct word from the brackets.

   A reaction or process that releases heat energy is described as (**exothermic/endothermic**).

   A reaction or process that takes in heat energy is described as (**exothermic/endothermic**).

   In combustion, a substance reacts with (**nitrogen/oxygen/carbon dioxide**) releasing energy.

2. Hydrocarbons such as methane burn in a plentiful supply of oxygen to release heat energy along with carbon dioxide and water.

   a Write a balanced chemical equation for this reaction.

   b State what is meant by the term hydrocarbon.

   c Suggest what might be produced when methane is burned in a limited supply of oxygen.

   d The products of combustion can be tested in the lab.

   Complete the diagram below to show how the products of combustion of methane could be tested.

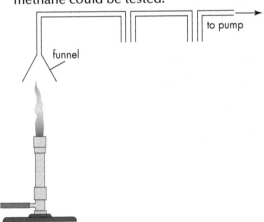

   to pump

   funnel

3. Calculate the energy required, in kJ, to heat:

   Hint $E_h = cm\Delta T$

   a 200 cm³ of water by 10 °C

   b 50 cm³ of water by 5 °C

   c 100 cm³ of water from 15 to 30 °C

   d 500 cm³ of water from 25 to 50 °C

   e 1 l of water by 15 °C

   f 200 cm³ of water from 20 to 60 °C.

4 Calculate the mass of water, in kg, that has been heated in each of the following:

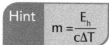

Hint $m = \dfrac{E_h}{c\Delta T}$

a 100 kJ of energy was released heating the water from 10 to 15 °C

b 500 kJ of energy was released heating the water from 20 to 90 °C

c 150 kJ of energy was released heating the water from 25 to 50 °C

d 200 kJ of energy was released heating the water from 20 to 40 °C

e 50 kJ of energy was released heating the water from 20 to 25 °C

f 300 kJ of energy was released heating the water from 20 to 40 °C

5 Calculate the increase in temperature, in °C, that would be recorded for each of the following:

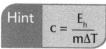

Hint $\Delta T = \dfrac{E_h}{cm}$

a 100 cm³ of water is heated by 1 kJ of energy

b 500 cm³ of water is heated by 20 kJ of energy

c 0·5 l of water is heated by 150 kJ of energy

d 25 kJ of energy is used to heat 150 cm³ of water

e 100 kJ of energy is used to heat 500 cm³ of water

f 50 kJ of energy is used to heat 500 cm³ of water

6 Calculate the specific capacity, in kJ kg⁻¹ °C⁻¹, of the following substances using the information given:

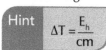

Hint $c = \dfrac{E_h}{m\Delta T}$

a 1·8 kJ of energy was required to heat 100 g of aluminium by 20 °C

b 55 kJ of energy was required to heat 1000 g of beryllium by 30 °C

c 81 kJ of energy was required to heat 700 g of copper by 300 °C

d 527 kJ of energy was required to heat 1·5 kg of silicon by 500 °C

e 48 kJ of energy was required to heat 500 g of ethanol from 20 °C to 40 °C

f 166 kJ of energy was required to heat 800 g of sand from 20 °C to 270 °C

7 The following experiment can be performed to establish the energy released on combustion of ethanol.

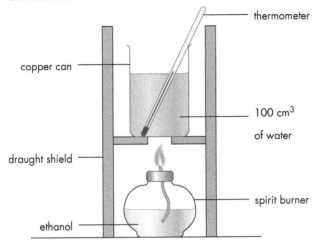

a **i** Draw the full structural formula of ethanol.

   ii Name the functional group present in ethanol.

b State all the measurements that would have to be recorded to calculate the energy released.

c It was found that burning 2 g of ethanol increased the temperature of the water by 40 °C.

   Calculate the energy released, in kJ, by ethanol.

d Suggest why the use of a draught shield and a copper can improves the results of this experiment.

e During the experiment, the copper can turns black due to the production of soot. Suggest why soot is produced.

3A Reactions of metals

1 Which of the following shows metallic bonding?

A 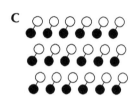   B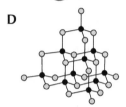

C   D

2 Which of the following would have the greatest rate of reaction?

> **Hint** The electrochemical series on page 10 of the SQA data booklet is very similar to the reactivity series and will help to answer these questions.

A 1 mol l^{-1} hydrochloric acid and 5 g of copper powder

B 1 mol l^{-1} hydrochloric acid and 5 g of copper lumps

C 1 mol l^{-1} hydrochloric acid and 5 g of zinc powder

D 1 mol l^{-1} hydrochloric acid and 5 g of zinc lumps

3 Which of the following would have the slowest rate of reaction?

A 1 mol l^{-1} hydrochloric acid and 5 g of tin powder

B 1 mol l^{-1} hydrochloric acid and 5 g of tin lumps

C 1 mol l^{-1} hydrochloric acid and 5 g of zinc powder

D 1 mol l^{-1} hydrochloric acid and 5 g of zinc lumps

4 The physical and chemical properties of a metal are due to the metallic bonds, which hold the atoms together.

a Explain fully how metallic bonds hold the atoms of a metallic element together.

b All metals conduct electricity.

Explain fully why metals can conduct electricity.

5 Metals react with oxygen to produce metal oxides.

> **Hint** metal + oxygen → metal oxide

a Complete and balance the equations below.

  i $Ca + O_2 \rightarrow$

  ii $Be + O_2 \rightarrow$

  iii $\rightarrow MgO$

  iv $\rightarrow Al_2O_3$

  v $+ O_2 \rightarrow Fe_2O_3$

b Which of the oxides shown would dissolve in water to form an alkaline solution?

6 Reactive metals react with water to produce metal hydroxides.

> Hint metal + water→ metal hydroxide + hydrogen

a Complete and balance the equations below.

  i $Ca(s) + H_2O(l) \rightarrow$

  ii $Na(s) + H_2O(l) \rightarrow$

  iii $ \rightarrow KOH(aq) + H_2(g)$

  iv $ \rightarrow Sr(OH)_2(aq) + H_2(g)$

  v $Li(s) + H_2O(l) \rightarrow$

b Write the ionic formulae of the metal hydroxides formed by each reaction.

7 Metals above hydrogen in the electrochemical series react with acids such as hydrochloric acid.

> Hint metal + acid → salt + hydrogen

a Complete and balance the equations below.

  i $Mg + HCl \rightarrow$

  ii $Ca + H_2SO_4 \rightarrow$

  iii $K + HNO_3 \rightarrow$

  iv $ \rightarrow Na_2SO_4 + H_2$

  v $Sr + HNO_3 \rightarrow$

b Write the ionic equations for the reactions.

8 The order of reactivity of metals can be compared by comparing their rate of reaction with oxygen as shown.

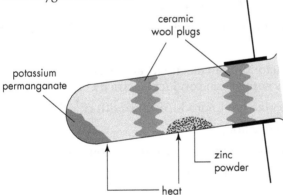

a The equation for the reaction taking place is:

$$Zn(s) + O_2(g) \rightarrow ZnO(s)$$

  i Balance the equation.

  ii Using the formula of zinc oxide, state the valency of zinc.

  iii Suggest why the zinc was used as a powder.

b The experiment was repeated and the results recorded.

Metal	Observation
Zinc	Glowed brightly
Magnesium	Bright white light produced
Copper	Dull red glow

  i Write the balanced chemical equation for the reaction between magnesium and oxygen.

  ii Use the results to place the metals in order of reactivity from the most to the least reactive.

9 Iron can be extracted from naturally occurring iron compounds in a blast furnace.

a State the term used to describe naturally occurring metal compounds.

b i Most naturally occurring iron compounds contain iron(III) ions which are reduced to produce the iron metal.

Write the ion-electron equation for this reaction.

ii The overall reaction that takes in the blast furnace is shown by the equation below:

$$Fe_2O_3 + 3CO \rightarrow 2Fe + 3CO_2$$

Calculate the mass, in grams, of iron produced when 1000 g of iron oxide reacts with carbon monoxide.

10 Copper is a very good conductor of both electricity and heat.

a Explain why metals such as copper can conduct electricity.

b Copper can react with chlorine gas to produce copper(II) chloride.

Write the equation for this reaction.

c Copper can be extracted from naturally occurring compounds such as malachite.

i State the term used to describe naturally occurring metal compounds such as malachite.

ii During the extraction of copper from its ore, copper ions are converted to copper atoms.

State the name of this type of reaction.

3B Redox

1 Complete the statements below by selecting the correct word from the brackets.

> **Hint**
> **Oil Rig**
> **Oxidation is loss**   **Reduction is gain**
> When substances lose electrons it is known as oxidation.
> When substances gain electrons it is known as reduction.

a Oxidation is the (**gain/loss**) of electrons by a reactant in any reaction.

b Reduction is the (**gain/loss**) of electrons by a reactant in any reaction.

c In a (**reduction/redox/oxidation**) reaction, both oxidation and reduction take place at the same time.

2 Write an ion-electron equation for each of the following reactions:

> **Hint** Refer to page 10 of the SQA data booklet for help if required.

a the oxidation of magnesium atoms

b the reduction of bromine atoms

c the oxidation of sodium atoms

d the reduction of fluorine atoms

e the reduction of copper(II) ions to copper atoms

f the oxidation of iodide ions to iodine atoms

g the reduction of tin(iv) ions to tin atoms

h the oxidation of phosphide ions

3 Complete the following ion-electron equations and label each as oxidation or reduction.

a $Ca \rightarrow Ca^{2+}$

b $F \rightarrow F^-$

c $Li \rightarrow Li^+$

d $Al \rightarrow Al^{3+}$

e $O \rightarrow O^{2-}$

f $N \rightarrow N^{3-}$

4 Name the reactant being oxidised in the following reactions.

a $Ca + MgSO_4 \rightarrow Mg + CaSO_4$

b $Cu(NO_3)_2 + Zn \rightarrow Zn(NO_3)_2 + Cu$

c $2I^- + 2Fe^{3+} \rightarrow I_2 + 2Fe^{2+}$

d $2AgNO_3 + Mg \rightarrow 2Ag + Mg(NO_3)_2$

5 For each of the following redox reactions write the ion-electron equations for both the reduction and oxidation steps.

a $Zn(s) + Cu^{2+}(aq) \rightarrow Zn^{2+}(aq) + Cu(s)$

b $3Mg(s) + 2Fe^{3+}(aq) \rightarrow 3Mg^{2+}(aq) + 2Fe(s)$

c $2KBr(aq) + Cl_2(g) \rightarrow Br_2(g) + 2KCl(aq)$

d $CaCl_2(aq) + F_2(g) \rightarrow CaF_2(aq) + Cl_2(g)$

6 Combine the following ion-electron equations to form a redox equation.

Hint	There must be an equal number of electrons in each equation before they can be combined to form the redox equation.

Reduction – $\quad Ag^+ + e^- \rightarrow Ag$ ← This equation must be multiplied by two so that there are an equal number of electrons in each equation.

Oxidation – $\quad Mg \rightarrow Mg^{2+} + 2e^-$

Reduction – $\quad 2Ag^+ + 2e^- \rightarrow 2Ag$

Oxidation – $\quad Mg \rightarrow Mg^{2+} + 2e^-$

Redox – $\quad 2Ag^+ + Mg \rightarrow Mg^{2+} + 2Ag$

a $Cu^{2+}(aq) + 2e^- \rightarrow Cu(s)$

$\quad Mg(s) \rightarrow Mg^{2+}(aq) + 2e^-$

b $Ag^+(aq) + e^- \rightarrow Ag(s)$

$\quad Mg(s) \rightarrow Mg^{2+}(aq) + 2e^-$

c $Zn^{2+}(aq) + 2e^- \rightarrow Zn(s)$

$\quad Al(s) \rightarrow Al^{3+}(aq) + 3e^-$

d $MnO_4^-(aq) + 5e^- + 8H^+(aq) \rightarrow Mn^{2+}(aq) + H_2O(l)$

$\quad Fe^{2+}(aq) \rightarrow Fe^{3+}(aq) + e^-$

e $Cu^{2+}(aq) + 2e^- \rightarrow Cu(s)$

$\quad Al(s) \rightarrow Al^{3+}(aq) + 3e^-$

f $Cr_2O_7^{2-}(aq) + 6e^- + 14H^+(aq) \rightarrow Cr^{3+}(aq) + 7H_2O(l)$

$\quad Sn^{2+}(aq) \rightarrow Sn^{4+}(aq) + 2e^-$

3C Extraction of metals

1 Which of the following metals is obtained from its ore by electrolysis?

> **Hint** The method used to extract a metal from its ore depends on the reactivity of the metal.

A Iron

B Copper

C Mercury

D Aluminium

2 Some metals can be obtained from their metal oxides by heat alone.

Which of the following oxides would produce a metal when heated?

A Calcium oxide

B Copper oxide

C Zinc oxide

D Silver oxide

3 Most metals have to be extracted from their ores. Place the following metals in the correct space in the table:

copper   mercury   aluminium

You may wish to use the SQA data booklet to help you.

Metal	Method of extraction
	using heat alone
	electrolysis of molten ore
	heating with carbon

4 During the extraction of metals, metal ions are reduced forming metal atoms.

Write the ion-electron equation for the formation of metal atoms from the following ores.

a Fe_2O_3

c Al_2O_3

d PbS

e CuO

f $ZnCO_3$

g SnO_2

5 Electrolysis of copper(II) chloride can be performed as shown.

> **Hint** Electrolysis is the breaking up of an ionic compound using electricity.

a State why a D.C. power supply must be used.

b Write the ion-electron equation for the reaction taking place at the negative electrode.

c Write the ion-electron equation for the reaction taking place at the positive electrode.

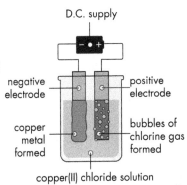

D.C. supply

negative electrode

positive electrode

copper metal formed

bubbles of chlorine gas formed

copper(II) chloride solution

6 Aluminium can be extracted from aluminium oxide as shown below.

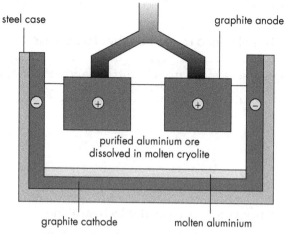

The aluminium oxide is dissolved in molten cryolite to lower the melting point of the aluminium oxide.

a State why the aluminium oxide must be molten to carry out electrolysis.

b Suggest why graphite is a suitable material to use as an electrode.

c Aluminum is produced as the negative electrode.

  i Write the ion-electron equation for the reaction taking place at the negative electrode.

  ii State the name for the reaction taking place at the negative electrode.

d Oxygen is produced at the positive electrode.

  i Write the ion-electron equation for the reaction taking place at the positive electrode.

  ii State the name for the reaction taking place at the positive electrode.

3D Electrochemical cells

Hint In an electrochemical cell, electrons travel from the most reactive metal to the least reactive metal through the wires. The further apart the metals are in the electrochemical series then the greater the voltage produced by the cell.

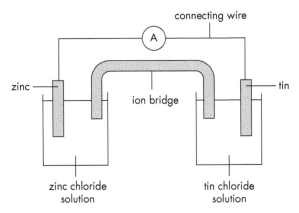

In the cell shown, electrons flow through:

A the wires from tin to zinc

B the wires from zinc to tin

C the ion-bridge from tin to zinc

D the ion-bridge from zinc to tin

2 Four cells were made by joining copper, iron, tin and zinc to silver.

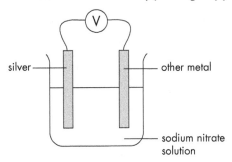

The voltages produced were recorded in the table shown.

Cell	Voltage (V)
A	1·5
B	1·1
C	0·8
D	0·4

Which line in the table shows the voltage of the cell containing zinc joined to silver?

3 Which pair of metals, when connected in a cell, would give the lowest voltage and a flow of electrons from **X** to **Y**?

(You may wish to use the SQA data booklet to help you.)

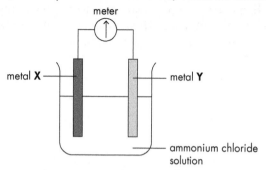

	Metal X	Metal Y
A	magnesium	copper
B	copper	magnesium
C	zinc	tin
D	tin	zinc

4 Four cells were made by joining copper, iron, magnesium and zinc to silver.

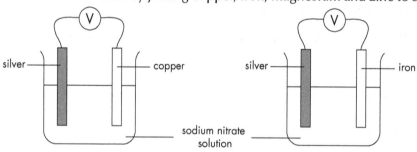

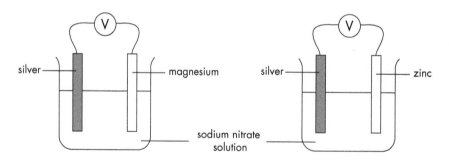

Which of the following will be the voltage of the cell containing magnesium joined to silver?

A 0·6 V   B 1·0 V

C 1·2 V   D 2·8 V

5 Iron displaces silver from silver(I) nitrate solution.

> **Hint** Spectator ions remain unchanged. They must have the same charge and be in the same state on both sides of the equation to be classified as spectator ions.

The equation for the reaction is:

$$Fe(s) + 2Ag^+(aq) + 2NO_3^-(aq) \rightarrow Fe^{2+}(aq) + 2Ag(s) + 2NO_3^-(aq)$$

a Identify the spectator ion in the reaction shown.

b Write the ion-electron equation for the **reduction** step in the reaction.

c This reaction can also be carried out in a cell.

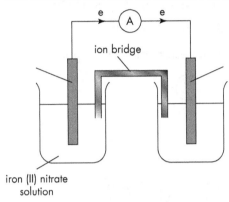

iron (II) nitrate solution

Complete the labels on the diagram.

d What is the purpose of the ion bridge?

6 Shown below is an electrochemical cell that does not involve metals.

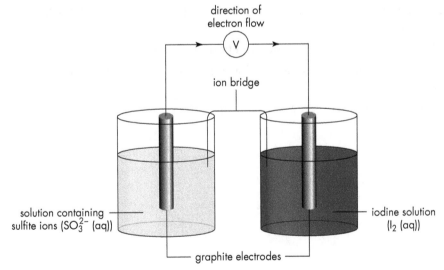

The equation for the reaction taking place in the beaker on the left can be represented by the equation:

$$SO_3^{2-}(aq) + H_2O(l) \rightarrow SO_4^{2-}(aq) + 2H^+ + 2e^-$$

a State the name of the type of reaction taking place in this beaker.

b State why graphite is a suitable material to use as an electrode.

c In the beaker containing iodine the iodine molecules are reduced to form iodide ions.

Write the ion-electron equation for this reaction.

d State the purpose of the ion bridge.

7 A cell can be produced by connecting iron(II) sulfate solution and acidified potassium permanganate solution. The iron(II) ions are oxidised and the permanganate ions are reduced.

The cell uses graphite electrodes, an ion bridge and a voltmeter.

a Draw and label a diagram of the cell including the direction of the flow of electrons.

b Write the ion-electron equation for the oxidation of Fe^{2+} ions to form Fe^{3+} ions.

You may wish to use the SQA data booklet.

c During the process the permanganate ions are reduced.

$$MnO_4^-(aq) + 8H^+(aq) + 6e^- \rightarrow Mn^{2+}(aq) + 4H_2O(l)$$

Write the overall redox reaction equation for this cell.

8 A student set up a cell by connecting magnesium metal to copper.

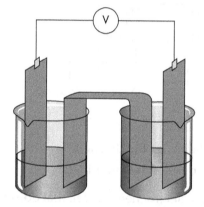

a Complete the diagram by labelling the metals and the electrolytes that could be used.

b Label the diagram showing the direction of electron flow.

c Write the ion-electron equations for both the oxidation and reduction reactions taking place.

d Write the overall redox reaction for this cell.

3E Plastics

1 Complete the statements below by selecting the correct word from the brackets.

a Plastics are examples of materials known as (**monomers/polymers**).

b (**Monomers/polymers**) are long chain molecules formed by joining together a large number of small molecules called (**monomers/polymers**).

c (**Addition/condensation**) polymerisation is the name given to a chemical reaction in which (**saturated/unsaturated**) monomers are joined, forming a polymer.

2 Complete the table below by adding the name of the monomer or polymer.

Monomer	Polymer produced
Ethene	
Butene	
	Polypropylene
	Poly(ethanol)
Urethane	
Vinyl chloride	
	Poly(tetrafluoroethene)

3 Part of the poly(chloroethene) chain is shown:

> **Hint** When drawing polymers always ensure that both end bonds are shown. This indicates that what you have drawn is only a section of a much larger chain.

a Draw the structure of the repeating unit.

b Name and draw the full structural formula of the monomer.

4 Complete the structural formulae and the repeating unit for each of the following:

Alkene monomer	Structural formula	Polymer	Structure of polymer (repeating unit)
ethene		poly(ethene)	
propene		poly(propene)	
phenylethene		poly(phenylethene) (polystyrene)	
tetrafluoroethene		poly(tetrafluoroethene) (Teflon®, PTFE)	

5 Polyacrylonitrile fibres can be used to produce knitted goods such as sweaters and socks. The monomer used to produce polyacrylonitrile is shown.

$$H_2C=CH-C\equiv N$$

  a Name the monomer used to make polyacrylonitrile.

  b Draw the structure of polyacrylonitrile showing three monomer units combined.

6 Poly(propene) is a polymer with many uses such as packaging. The structure of the monomer used to produce poly(propene) is shown.

$$H_3C-CH=CH_2$$

  a Draw the structure of a poly(propene) molecule showing three monomer units combined.

  b Draw the structure of the repeating unit of poly(propene).

  c Name the type of polymerisation that takes place to form poly(propene).

7 Shown below is the structure of the polymer Perspex.

  a Draw the structure of the repeating unit of Perspex.

  b Draw the full structural formula of the monomer used to produce Perspex.

3F Fertilisers

1 Fertilisers provide the essential elements required for healthy plant growth.

  a Name the three elements that are essential for healthy plant growth.

  b Which of the following properties must all fertilisers have?

   i Flammable

   ii Soluble in water

   iii pH of 7

   iv Boiling point of over 150 °C

   v Conducts electricity

2 Listed below are three compounds that are commonly used as fertilisers.

Compound name	Solubility in water
Ammonium phosphate	Very soluble
Potassium nitrate	Very soluble
Ammonium sulfate	Very soluble

  a Write the chemical formula of each compound.

  b Write the formula of each compound showing the charges on each ion.

  c Which of the compounds provides only one of the elements essential for healthy plant growth?

  d Suggest why fertilisers must be soluble in water.

  e Suggest a test that could be performed to show that potassium nitrate contains potassium.

3 The use of the essential elements, N, P and K, has increased steadily.

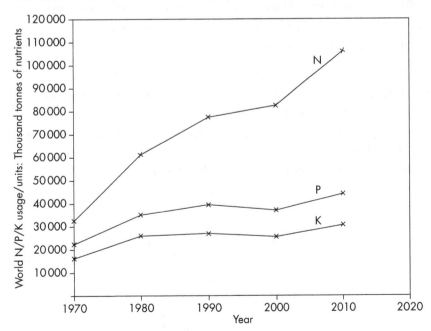

Source: International Fertilizer Industry Association (IFA)

  a Estimate how many million tonnes of N, P and K were used in 2015.

  b Suggest why the amount of N, P and K used is increasing.

4 Ammonia and nitric acid can be used to produce soluble, nitrogen-containing salts.

> **Hint** It is important to learn the properties of ammonia and the various ways it can be produced and collected.

a Write the formulae of both ammonia and nitric acid.

b Ammonia reacts with nitric acid to produce ammonium nitrate.

  i Name the type of reaction that takes place between the ammonia and nitric acid.

  ii Write the formula equation for the reaction.

c State a use for the ammonium nitrate produced.

5 Ammonia can be produced in the lab by performing the experiment shown.

Ammonium chloride is mixed with sodium hydroxide solution and heated, which produces sodium chloride, water and ammonia gas.

a Copy and complete the diagram by adding the labels.

b Write the formula equation for the reaction.

c Moist pH paper can be placed at the mouth of the test tube to test for ammonia gas. State the colour that the pH paper would turn if ammonia gas is produced.

d In industry ammonia is produced using another process. Name the industrial process used to produce ammonia.

6 Ammonia can be converted to nitric acid by the Ostwald process.

a Name the other two reactants that are required in the Ostwald process.

b Name the catalyst used in the Ostwald process.

c The final stage involves the dissolving of nitrogen dioxide in water. Write the equation for this reaction.

d Complete the flow diagram of the Ostwald process.

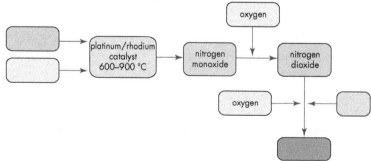

7 Ammonia is an important compound in the manufacture of fertilisers and explosives.

a i Draw a diagram of an ammonia molecule showing all the outer electrons.

ii Draw a diagram showing the shape of an ammonia molecule.

iii Name the shape of an ammonia molecule.

b Ammonia solution reacts with acids to form salts.

Name the salts produced when an ammonia solution reacts with the following acids:

i sulfuric acid

ii nitric acid

iii hydrochloric acid

iv Write the equations for each of the neutralisation reactions that take place between the acids and ammonia solution.

c Ammonia is produced industrially by the Haber process, which uses an iron catalyst that allows the experiment to be performed at lower temperatures, making the process more economical.

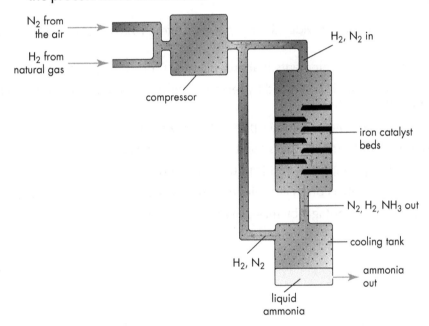

i Suggest why the iron catalyst is used in the powdered form.

ii Describe another part of the process that makes the Haber process more economical.

iii Write the balanced chemical equation for the Haber process.

d Along with the iron catalyst a moderately high temperature is used.

i Explain fully why a high temperature is not used in the Haber process.

ii Explain fully why a low temperature is not used in the Haber process.

8 An investigation was performed into the best temperature and pressure to use in the Haber process. The first experiment changed the pressure used and recorded the yield of ammonia produced.

Pressure (atmospheres)	100	200	300	400	500
Percentage yield (%)	10	20	30	35	40

a Predict the percentage yield of ammonia produced at a pressure of 600 atmospheres.

b The second experiment changed the temperature used and recorded the yield of ammonia produced.

Temperature (°C)	200	300	400	500
Percentage yield (%)	90	65	40	15

Suggest why a temperature of 600 °C would not be used.

c Which combination of temperature and pressure would produce the highest yield of ammonia?

9 **a** Ammonium nitrate provides the element nitrogen, which is essential for healthy plant growth.

Name the two other elements that are essential for healthy plant growth.

b Ammonium nitrate is produced by reacting ammonia solution with nitric acid in a neutralisation reaction.

$$NH_3 \text{ (aq)} + HNO_3\text{(aq)} \rightarrow NH_4NO_3\text{(aq)} + H_2O \text{ (l)}$$

i Name the industrial process used to make ammonia.

ii Name the spectator ion in this neutralisation reaction.

c Ammonium nitrate is often combined with urea, $CO(NH_2)_2$, for use as a fertiliser.

Calculate the percentage, by mass, of nitrogen in urea.

3G Radioactivity

1 Complete the statements below by selecting the correct word from the brackets.

   a Radioactive decay involves changes in the (**nuclei/energy**) levels of atoms.

   b Unstable (**nuclei/atoms**) (radioisotopes) can become more stable by giving out
(**alpha/beta/gamma**) (α), (**alpha/beta/gamma**) (β) or (**alpha/beta/gamma**) (γ) radiation.

2 **a** Complete the following table.

Radiation	Alpha	Beta	Gamma
Symbol			
Mass			
Charge			

   b Identify each of the types of radiation, **X**, **Y** and **Z**, shown in the diagram below.

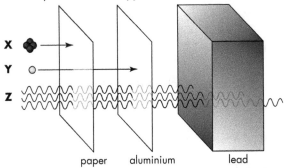

3 Alpha, beta and gamma radiation are passed through charged plates.

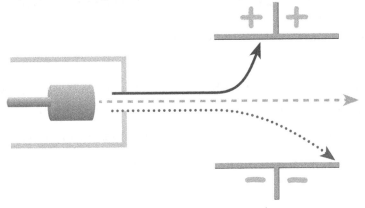

   a State what type of radiation is shown by the solid line and explain your answer.

   b State what type of radiation is shown by the dashed line and explain your answer.

   c State what type of radiation is shown by the dotted line and explain your answer.

4 Write the balanced nuclear equation for each of the following alpha-emitting isotopes:

> **Hint** In nuclear equations both the mass numbers and atomic numbers on each side must be equal, e.g.
>
>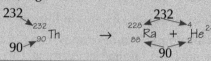

a $^{241}_{95}Am$ b $^{210}_{84}Po$

c $^{237}_{93}Np$ d $^{249}_{98}Cf$

e $^{209}_{83}Bi$ f $^{215}_{85}At$

5 Write the balanced nuclear equation for each of the following beta-emitting isotopes:

a $^{66}_{29}Cu$ b $^{209}_{82}Pb$

c $^{66}_{28}Ni$ d $^{90}_{39}Y$

e $^{188}_{74}W$ f $^{90}_{38}Sr$

6 Write the balanced nuclear equation for each of the following proton-emitting isotopes:

a $^{14}_{8}O$ b $^{13}_{7}N$

c $^{22}_{11}Na$ d $^{82}_{37}Rb$

e $^{52}_{26}Fe$ f $^{121}_{53}I$

7 When beryllium-9 is combined with an alpha particle, a neutron is produced along with a carbon-12 atom.

a Complete the equation for the reaction described above.

$^{9}_{4}Be$ +   →   + $^{12}_{6}C$

b Along with the neutron and carbon-12 atom, gamma radiation is also produced. State why gamma radiation does not have an atomic or mass number.

8 An atom of $^{227}_{90}Th$ decays by a series of alpha emissions to form an atom of $^{211}_{82}Pb$.

Calculate the number of alpha particles released by this process.

9 Americium-241, a radioisotope used in household smoke detectors, is an alpha-emitting isotope.

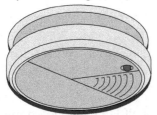

a What is meant by the term isotope?

b Write a balanced nuclear equation for the alpha decay of $^{241}_{95}Am$.

c Americium-241 has a half-life of 433 years. State what is meant by the term half-life.

10 Radium was used in the early 1900s to make clock faces and dials glow in the dark. The most common isotope of radium is radium-226, which has a half-life of 1600 years.

a The equation for the decay of radium-226 is:

$$^{226}_{88}Ra \rightarrow \mathbf{X} + ^{4}_{2}He$$

Name element **X**.

b Name the type of radiation emitted by the radium-226 isotope.

3H Half-life

1 The half-life of lead-210 is 21 years.

  a State what is meant by the term half-life.

  b What fraction of the original lead-210 isotope would remain after 63 years?

2 Iodine-131 is a radioisotope with a half-life of 8 days, which can be used in the treatment of certain cancers.

  a Suggest why isotopes with a short half-life are used in medicine.

  b Calculate the fraction of iodine-131 that would remain after 32 days.

  c The amount of potassium iodide prescribed to children over the age of 12 is 0·13 g.

   Calculate the number of moles of potassium iodide prescribed.

3 The decay curve for a 1·0 g sample of technetium-99 is shown.

  a Using the graph, calculate the half-life of technetium-99.

  b Technetium-99 can be used to detect heart problems by injecting it into the body, despite it being a gamma-emitting radioisotope.

   Suggest why technetium-99 can be safely used in this way.

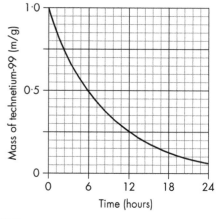

4 The decay curve for a sample of a radioisotope is shown.

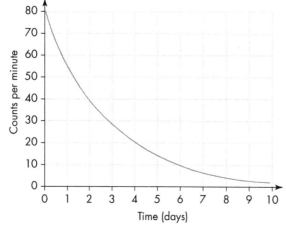

  a State the half-life of the sample.

  b Explain what effect increasing the temperature would have on the half-life of the radioisotope.

5 Phosphorus-32 can be used to detect cancerous tumours in the body. The decay curve for a sample of phosphorus-32 is shown.

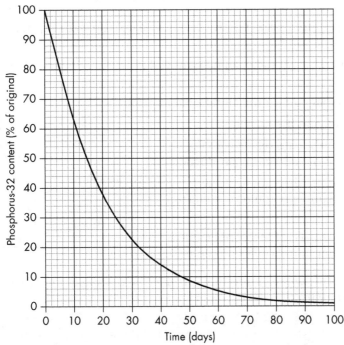

a Using the graph, calculate the half-life of phosphorus-32.

b State what effect increasing the temperature would have on the half-life of phosphorus-32.

c Calculate the time taken for a 100 g sample of phosphorus-32 to decrease to 6·25 g.

d Phosphorous-32 decays by beta emission.

Write the equation for the beta decay of phosphorous-32.

6 Polonium-210 has a half-life of 140 days and decays by alpha emission.

a Draw a line graph on graph paper to show how the mass of 100 g of polonium-210 would change over time.

b Write the equation for the alpha decay of polonium-210.

7 Americium-241 is an alpha-emitting isotope with a half-life of 432 years and is used in household smoke detectors.

a Write the equation for the decay of americium-241.

b Suggest why smoke detectors containing americium-241 are safe to use in homes.

c Suggest why a radioisotope with a long half-life is required in smoke detectors.

General Practical Techniques

1 The soluble salt copper sulfate can be produced by reacting excess copper carbonate with sulfuric acid.

a The unreacted copper carbonate can be removed from the copper sulfate solution by filtration.

   i Draw a labelled diagram for this filtration.

   ii Name the residue and the filtrate.

b When the copper carbonate reacts with the sulfuric acid, carbon dioxide gas is produced.

Copy and complete the diagram to show how the carbon dioxide gas could be collected and measured.

c Suggest how a dry sample of the copper sulfate can be obtained when filtration is complete.

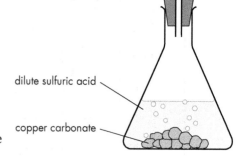

dilute sulfuric acid

copper carbonate

2 The table below lists various gases and their properties. Complete the table to show what method should be used to collect each gas.

Gas	Solubility in water	Density (compared to air)	Method used to collect gas
Ammonia	High	Low	
Hydrogen chloride	High	High	
Methane	Low	Low	

3 Lead(II) iodide can be produced by reacting lead(II) nitrate solution with potassium iodide solution.

a Write the balanced formula equation, including state symbols, for this reaction.

You may wish to use the SQA data booklet.

b Suggest how a dry sample of lead iodide could be obtained when the reaction is complete.

4 Flame tests can be used to identify the presence of various metal ions.

Complete the table below.

You may wish to use the SQA data booklet.

Flame colour	Metal responsible
Yellow	
Lilac	
Green	
Blue-green	
	Lithium
	Calcium
	Strontium

5 Ammonia is a base that reacts with nitric acid to produce the salt ammonium nitrate. Ammonia and nitric acid react in a ratio of 1:1.

A titration is performed to establish the concentration of ammonia solution by performing a titration with a 2·0 mol l⁻¹ standard solution of nitric acid. The results are recorded in the table below.

Titration	Volume added (cm³)
1	9·8
2	10·1
3	10·1
4	10·2

a Calculate the average volume of nitric acid, in cm³, that should be used to calculate the concentration of the ammonia solution.

b State what must be added to the flask to show the end-point of the reaction.

c 20 cm³ of the ammonia solution was transferred to the conical flask used in the titration.

State the name of the piece of equipment that should be used to accurately measure the 20 cm³ of ammonia solution.

d Calculate the concentration of the ammonia solution.

Your answer must include the appropriate unit.

6 Chlorine gas can be produced by reacting concentrated hydrochloric acid with manganese dioxide. The chlorine gas produced is first dried by bubbling it through concentrated sulfuric acid before collecting the gas by the upward displacement of air.

a Complete the diagram.

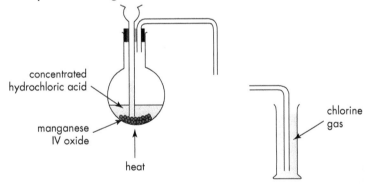

b State what this method of collecting chlorine gas suggests about its density relative to air.

A student was testing the products released on the combustion of ethanol.

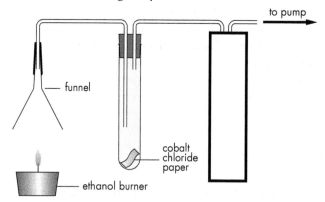

a Complete the diagram to show how the student could have tested for the production of carbon dioxide.

b The balanced equation for the combustion of ethanol is:

$$C_2H_5OH + 3.5O_2 \rightarrow 2CO_2 + 3H_2O$$

Calculate the mass, in g, of carbon dioxide that is released on combustion of 4·6 g of ethanol. Show your working clearly.

c Energy is also released on combustion of ethanol. State the name given to reactions that release energy.

8 A student carried out titrations to establish the concentration of a sodium hydroxide solution.

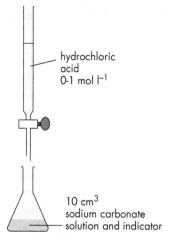

hydrochloric acid
0·1 mol l^{-1}

10 cm^3
sodium carbonate
solution and indicator

The results of the titration are given in the table below:

Titration	Initial burette reading (cm³)	Final burette reading (cm³)	Total (cm³)
1	0·0	15·9	15·9
2	15·9	31·0	15·1
3	31·0	45·9	14·9

a Calculate the average volume of acid, in cm³, that should be used in calculating the concentration of the sodium hydroxide solution.

b The equation for the reaction is:

$$HCl(aq) + NaOH(aq) \rightarrow NaCl(aq) + H_2O(l)$$

Calculate the concentration, in mol l^{-1}, of the sodium hydroxide solution.

Leckie
the education publisher
for Scotland

National 5
CHEMISTRY

Mixed Exam
Question Practice

Maria D'Arcy, Bob Wilson, Graeme Wilson

MULTIPLE-CHOICE QUESTIONS

1 Magnesium reacts with hydrochloric acid to form hydrogen gas. The speed of the reaction can be measured by recording the volume of hydrogen gas produced per minute.

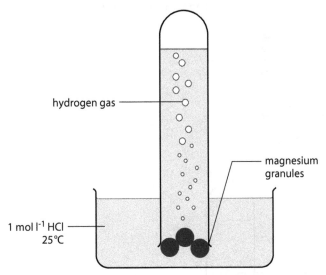

hydrogen gas

magnesium granules

1 mol l^{-1} HCl
25°C

Which experiment would you use to investigate the effect of temperature on the above reaction?

A

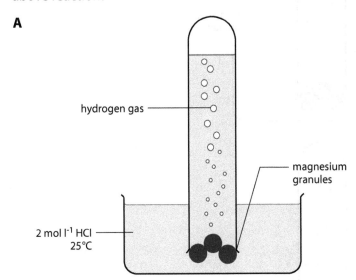

hydrogen gas

magnesium granules

2 mol l^{-1} HCl
25°C

B

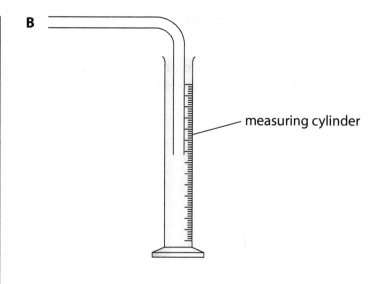

measuring cylinder

C

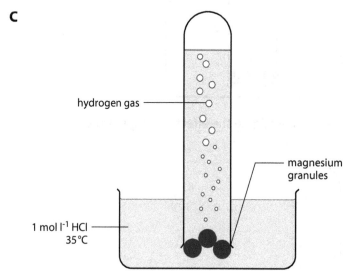

hydrogen gas

magnesium
granules

1 mol l^{-1} HCl
35 °C

D

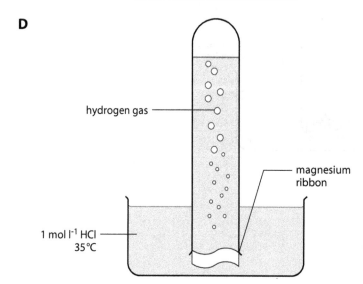

hydrogen gas

magnesium
ribbon

1 mol l^{-1} HCl
35 °C

1

2 Which set of information describes a proton?

Answer	Mass	Charge	Location
A	1	+1	orbitals/shells
B	1	0	nucleus
C	0	−1	orbitals/shells
D	1	+1	nucleus

1

3 $^{35}_{17}Cl$ and $^{37}_{17}Cl$ are

A isomers

B isotopes

C ions

D homologous series.

1

4 Which of the following elements is in the same period as Lithium?

You may wish to use the SQA data booklet to help you.

A Carbon

B Potassium

C Helium

D Magnesium

1

5 An unknown substance conducts electricity as a solid and as a liquid and is a liquid at room temperature.

The substance could be

A Bromine

B Water

C Magnesium

D Mercury

1

6 The following apparatus was set up.

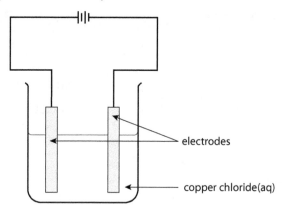

Which set of results matches the above electrolysis?

Answer	Positive	Negative
A	Solid copper	Chorine gas
B	Hydrogen gas	Chlorine gas
C	Chlorine gas	Solid copper
D	Hydrogen gas	Solid copper

1

7 Hydroxide ions are released when solid barium hydroxide is added to water.

What would be the pH of the resultant solution?

A 6

B 9

C 7

D 6·5

1

8 Which line in the table describes what happens when water is added to an acid?

	pH	H+ ion concentration
A	increases	increases
B	decreases	decreases
C	increases	decreases
D	decreases	increases

1

9 25 cm^3 of $Ca(OH)_2$ neutralised 20 cm^3 of $0\cdot1 \text{ mol l}^{-1}$ solution of hydrochloric acid, HCl.

$$Ca(OH)_2 + 2HCl \longrightarrow CaCl_2 + 2H_2O$$

What was the concentration of the calcium hydroxide, $Ca(OH)_2(aq)$?

You may wish to use the SQA data booklet.

A $0\cdot1 \text{ mol l}^{-1}$

B $0\cdot04 \text{ mol l}^{-1}$

C $0\cdot16 \text{ mol l}^{-1}$

D $0\cdot06 \text{ mol l}^{-1}$

1

10 The hydrocarbon below belongs to the cycloalkanes.

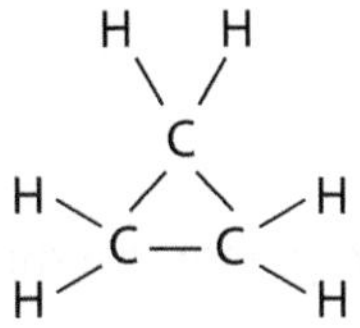

What is the general formula of the cycloalkanes?

A C_nH_n

B C_nH_{2n-2}

C C_nH_{2n}

D C_nH_{2n+2}

1

11 Which of the following hydrocarbons could be described as saturated?

A C_2H_2

B C_2H_4

C C_2H_6

D C_3H_4

1

12 Which of the following compounds is a solid at $0°C$?

You may wish to use the SQA data booklet to help you.

A Ethanol

B Ethanoic acid

C Propan-1-ol

D Propanoic acid

1

13 A compound which contains the functional group $-C=O$ is an

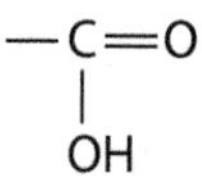

                                                      OH

A Acid

B Alcohol

C Alkali

D Alkene　　　　　　　　　　　　　　　　　　　　　　　　　　　　　　　1

14 Which term best describes the burning of ethanol?

A endothermic

B exothermic

C reduction

D redox　　　　　　　　　　　　　　　　　　　　　　　　　　　　　　　1

15 Which of the following structures represents a molecule which is the monomer used to make poly(propene)?

A

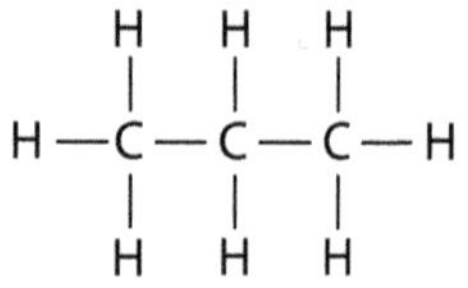

B

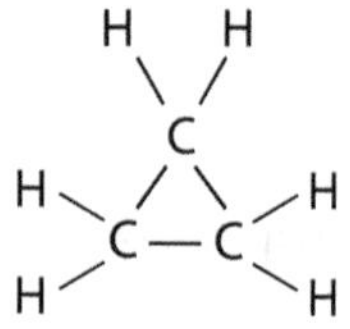

C

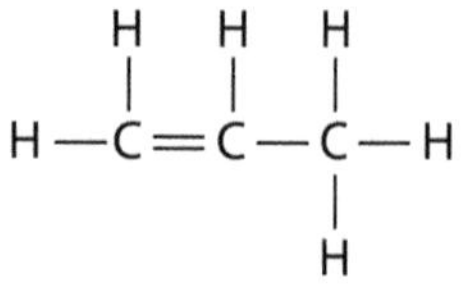

D

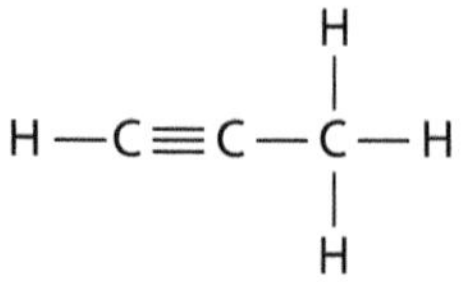

　　　　　　　　　　　　　　　　　　　　　　　　　　　　　　　　　　　1

16 The ion electron equations for the redox reaction between zinc and silver ions are shown below.

$$Zn(s) \longrightarrow Zn^{2+}(aq) + 2e$$

$$Ag^+(aq) + e \longrightarrow Ag(s)$$

The correct balanced redox equation is

A $Zn(s) + Ag^+(aq) \longrightarrow Zn^{2+}(aq) + Ag(s)$

B $2Zn(s) + Ag^+(aq) \longrightarrow 2Zn^{2+}(aq) + Ag(s)$

C $Zn(s) + 2Ag^+(aq) \longrightarrow Zn^{2+}(aq) + 2Ag(s)$

D $Zn(s) + Ag^+(aq) + e \longrightarrow Zn^{2+}(aq) + Ag(s) + 2e$

1

17 Which metal will not react with sulfuric acid?

A Silver

B Magnesium

C Aluminium

D Zinc

1

18 Which of the following metals can only be extracted from its oxide by electrolysis?

A Aluminium

B Copper

C Gold

D Zinc

1

19 Which of the following diagrams represents metallic bonding?

A

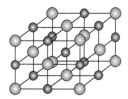

B

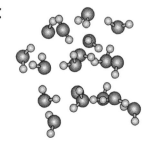

C

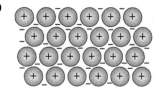

D

20 Which statement correctly describes what happens during Beta decay?

A An electron is fired out of the nucleus.

B Two protons and two neutrons are fired out of the nucleus.

C A neutron is fired out of the nucleus.

D An electromagnetic wave is released from the nucleus.

21 The chemical equation for the reaction between silver(I) nitrate and sodium chloride is:

$$AgNO_3 \text{ (aq)} + NaCl \text{ (aq)} \longrightarrow AgCl \text{ (s)} + NaNO_3 \text{ (aq)}$$

The best way to separate out the silver chloride is by

A   Distillation

B   Evaporation

C   Filtration

D   Precipitation

1

22 A sample of a compound containing a metal ion gave an orange-red flame colour when held in a hot Bunsen flame.

The metal could be

A   Barium

B   Calcium

C   Potassium

D   Sodium

1

23 Which of the following diagrams shows the most appropriate apparatus for collecting a sample of a soluble gas which is less dense than air?

A

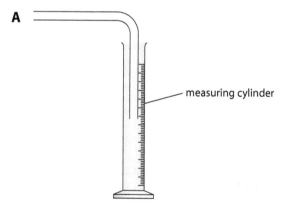

measuring cylinder

B

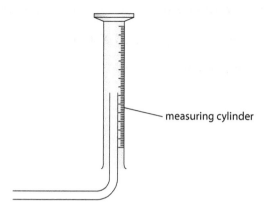

measuring cylinder

C

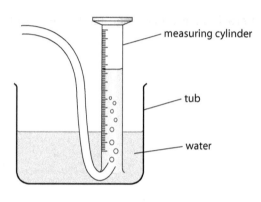

measuring cylinder

tub

water

D

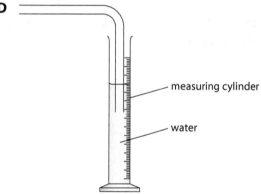

measuring cylinder

water

1

24 Which piece of chemical apparatus is used in a titration to accurately measure the volume of acid required to neutralise an alkali?

A Burette

B Dropper

C Measuring cylinder

D Pipette

1

25 Four titrations were carried out to find the volume of acid needed to neutralise an alkali.

The table shows the volumes of acid recorded at the end-point of each titration.

Titration	Volume of acid added (cm^3)
1 (rough)	12·9
2	12·7
3	12·3
4	12·1

Which volume of acid should be used in a calculation to find out the concentration of the acid?

A   12·9 cm^3

B   12·8 cm^3

C   12·5 cm^3

D   12·2 cm^3

1

Multiple-choice questions

EXTENDED ANSWER QUESTIONS

26 Zinc reacts with hydrochloric acid to produce hydrogen gas.

The table below shows the volume of hydrogen gas collected over time for the above reaction.

Time (seconds)	Volume of hydrogen (cm^3)
0	0
20	8
40	15
60	21
80	26
100	30
120	34
140	37
160	39
180	40
200	40

a **i** Plot a line graph of the results of the reaction.

3

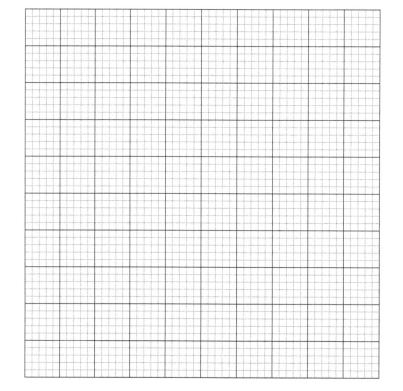

ii Calculate the average rate of reaction between 10 seconds and 40 seconds.

Your answer must include the appropriate unit.

Show your working clearly.

b This reaction can be increased by the addition of the catalyst copper.

i The graph below represents the volume of hydrogen gas evolved over time for the reaction between zinc and acid.

Draw the curve you would expect if the same experiment was repeated with the addition of copper.

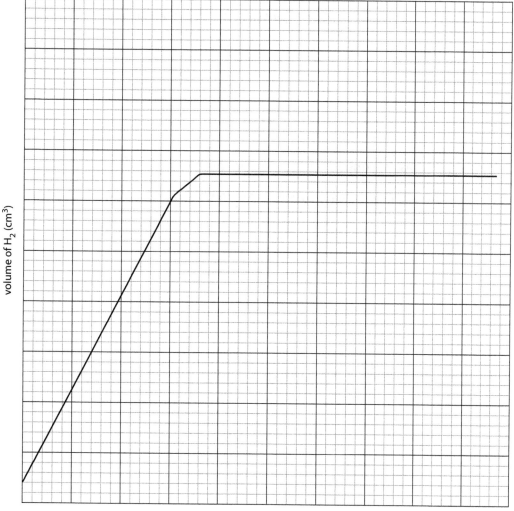

ii Draw and label the apparatus and chemicals you would use to determine the volume of hydrogen gas given off when zinc and hydrochloric acid react in the presence of copper.

27 Fuels made from plants are called biofuels.

Ethanol is a biofuel formed from the fermentation of carbohydrates.

The combustion of ethanol is shown below.

$$C_2H_5OH(l) + O_2(g) \longrightarrow CO_2(g) + H_2O(l)$$

a **i** Balance the above equation.

1

ii Calculate the mass of oxygen required to burn 9·2 g of ethanol.

3

b Methane (biogas) is formed from rotting plant matter.

i Draw the outer electrons in a molecule of methane, CH_4.

1

ii What name is given to the shape a molecule of methane takes?

1

28 Carbon forms many different compounds. Two such compounds are silicon carbide (SiC) and carbon dioxide.

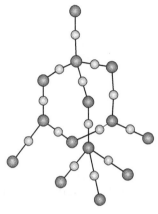

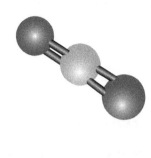

Silicon carbide   Carbon dioxide

The way in which the atoms are arranged (structure) determines their physical properties.

Carbon dioxide melts at $-57°C$ while silicon carbide has a melting point of $2730°C$.

a   **i** What type of structure does carbon dioxide take?

1

   ii In terms of bonding and structure explain the difference in melting points of the two compounds of carbon.

2

29 Calcium forms many useful products.

One such compound is calcium phosphate which is the main constituent of your skeleton.

a Write the formula for calcium phosphate.

b When calcium reacts it forms the ion Ca^{2+}.

    i In terms of electrons, describe the changes in a calcium atom when it reacts to form a calcium ion.

    ii Explain why calcium forms Ca^{2+} ions and not Ca^{3+} ions.

c Calcium (II) chloride is another compound of calcium and forms the following structure when solid.

What type of structure is shown below?

calcium (II) chloride

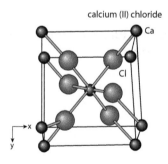

Extended answer questions

30 A student was measuring the pH of a number of solutions and concluded that only acids contain hydrogen ions and only alkalis contain hydroxide ions.

Using your knowledge of chemistry, comment on the student's conclusion.

31 The hydrogen halides when dissolved in water form a group of strong acids.

When the hydrogen halides dissolve in water, the bond between the hydrogen and the halide is broken and a hydrogen ion is released.

$$HCl(g) \longrightarrow H^+(aq) + Cl^-(aq)$$

The energy required to break the bond between the hydrogen and the halide is called its bond enthalpy.

Hydrogen halide	GFM	Bond enthalpy (kJmol l^{-1})
HF	20	+569
HCl	36·5	+428
HBr	81	+362
HI	128	

a i Write a statement linking the GFM of the halide to its bond enthalpy.

You may wish to use the SQA data booklet.

1

ii Predict the bond enthalpy of hydrogen iodide (HI).

1

b Hydrochloric acid reacts with the alkali calcium hydroxide. The reaction is given.

$$2H^+Cl^-(aq) + Ca^{2+}(OH^-)_2(aq) \longrightarrow Ca^{2+}Cl^-_2(aq) + 2H_2O(l)$$

i Circle the salt formed in the above reaction.

1

ii Rewrite the above reaction omitting all spectator ions.

1

32 Ethylchloride is a member of a group of compounds named the chloroalkanes and can be produced by reacting ethene with hydrogen chloride.

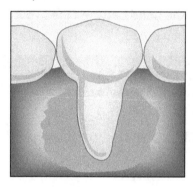

Ethylchloride is used in dentistry to detect dead teeth. A small amount of ethylchloride is sprayed onto a tooth. If the tooth is still alive the patient will feel a chilling effect due to the low boiling point of ethylchloride.

The following table shows the boiling points of some chloroalkanes.

Systematic name	Full structural formula	Boiling point (°C)
1-chloroethane		12
2-chloropropane		36
2-chlorobutane		70
3-chloropentane		96

a **i** Draw the full structural formula of the alkene that could be reacted with hydrogen chloride to make 3-chloropentane.

1

ii Name the type of chemical reaction which takes place when alkenes react with hydrogen chloride.

1

b Propene reacts with hydrogen in a similar way.

i Name the product of the reaction between propene and hydrogen.

1

ii Cyclopropane and propene are isomers of each other.

State what is meant by the term isomer.

1

iii When bromine solution is added to each hydrocarbon the following colour changes take place.

Colour change	Hydrocarbon
No change	
Yellow to clear	

Complete the table to show which hydrocarbon, cyclopropane and propene, give which result when added to bromine solution.

1

33 Alcohols are often found in hand gels. Alcohol-based hand sanitisers are more effective at killing bacteria than soaps and do not dry out skin as much.

The main ingredient in hand gels is

$$H-\underset{\underset{H}{|}}{\overset{\overset{H}{|}}{C}}-\underset{\underset{H}{|}}{\overset{\overset{OH}{|}}{C}}-\underset{\underset{H}{|}}{\overset{\overset{H}{|}}{C}}-H$$

a **i** Name the alcohol used in hand gels.

1

ii What name is given to the functional group present in all alcohols?

1

b The alcohol in (a) was formed from an alkene.

$$H-\underset{\underset{H}{|}}{\overset{\overset{H}{|}}{C}}-\underset{H}{\overset{H}{\underset{|}{C}}}=\underset{\underset{H}{|}}{\overset{H}{C}} + H_2O \longrightarrow H-\underset{\underset{H}{|}}{\overset{\overset{H}{|}}{C}}-\underset{\underset{H}{|}}{\overset{\overset{OH}{|}}{C}}-\underset{\underset{H}{|}}{\overset{\overset{H}{|}}{C}}-H$$

Two alcohols were formed, one of which is shown.

Draw the full structural formula of the other alcohol.

1

34 Ethanoic acid is a carboxylic acid. Vinegar is a solution of ethanoic acid and can be used in household cleaning products.

a Give one reason that vinegar can be used safely in cleaning products.

1

b Draw the structural formula for ethanoic acid.

1

c Name the functional group found in all carboxylic acids.

1

d The table below gives the boiling points of some carboxylic acids.

Carboxylic acid	Boiling point (°C)
ethanoic acid	118
propanoic acid	141
butanoic acid	164
pentanoic acid	

1

i Predict the boiling point of pentanoic acid.

ii Explain why the boiling points of the carboxylic acids increase as the size of the molecules increase.

1

e Name the salt formed when propanoic acid reacts with magnesium hydroxide.

1

35 A student was investigating the energy produced when 1 g of an alcohol burns. The work card the student followed is shown below. Two of the steps have been omitted.

Energy Work Card

Step 1: Measure out 100 cm³ of water into a copper can.

Step 2:

Step 3: Weigh the alcohol burner.

Step 4: Place the burner under the can and light it.

Step 5: After two minutes, put the flame out and measure the temperature of the water.

Step 6:

Step 7: Work out the change in the water's temperature and the mass of alcohol burned.

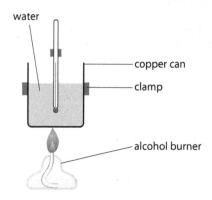

a Complete steps 2 and 6:   2

   Step 2:

   Step 6:

b Suggest why a copper can is used instead of a glass beaker.

c Suggest **one** improvement to the student's experimental set-up to ensure as much heat as possible is transferred to the water.

d i The temperature of the water rose by 11 °C.

Calculate the amount of energy produced, in kJ.

You may wish to use the SQA data booklet to help you.

Show your working clearly.

ii The mass of alcohol burned to produce the energy you calculated in part (i) is 0·22 g. Calculate the amount of energy, in kJg^{-1}, that would be produced when 1·0 g of the alcohol was burned.

Show your working clearly.

36 Iron is extracted from iron ore industrially in a blast furnace.

In a blast furnace, iron oxide is reacted with carbon dioxide to produce iron metal and carbon dioxide.

$$Fe_2O_3(s) + 3CO(g) \longrightarrow 2Fe(s) + 3CO_2(g)$$

a What is the charge on the iron ion in the iron oxide?　　　　　　　　　　1

b Write an ion electron equation for the reduction of iron in a blast furnace.　　1

c Iron is used to make large structures such as bridges and buildings.

Explain how iron structures can be damaged by acid rain.　　　　　　　1

37 Read the following passage carefully and answer the questions which follow.

Rechargeable batteries

An entirely new type of material developed at Rensselaer Polytechnic Institute could enable the next generation of high-power rechargeable lithium-ion (Li-ion) batteries for electric automobiles, as well as batteries for laptop computers, mobile phones and other portable devices.

The positive electrode (anode) in a Li-ion battery physically grows and shrinks as the battery charges or discharges, when in use. When charging the addition of Li-ions increases the volume of the electrode, while discharging has the opposite effect. These volume changes result in a build-up of stress in the anode. Too great a stress that builds up too quickly, as in the case of a battery charging or discharging at high speeds, can cause the battery to fail prematurely. This is why most batteries in today's portable electronic devices like cell phones and laptops charge very slowly – the slow charge rate is intentional and designed to protect the battery from stress-induced damage.

The Rensselaer team's nanoscoop positive electrode, however, was engineered to withstand this build-up of stress. Made from a carbon base topped with a thin layer of aluminum (Al) and a 'scoop' of silicon (Si), the structures are flexible and able to quickly accept and discharge Li-ions at extremely fast rates without sustaining significant damage.

The segmented structure of the nanoscoop allows the strain to be gradually transferred from the C base to the Al layer, and finally to the Si scoop. This natural strain gradation provides for a less abrupt transition in stress across the material interfaces, leading to improved structural integrity of the electrode.

This passage was adapted from the article 'Nanoscoops could spark new generation of electric automobile batteries' by Rensselaer Polytechnic Institute, published on the science daily.com, January 2011.

a Name **one** of the elements used to make the 'nanoscoop'.

1

b Oxidation and reduction reactions take place at the positive electrode (anode) depending on whether you are charging your cell or it is discharging and powering an appliance.

Describe the difference between an oxidation and reduction reaction.

1

c Why is the new electrode not made of lithium metal?

1

38 The diagram shows a cell made without using metals.

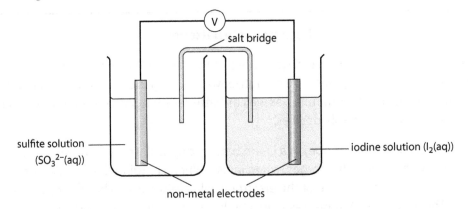

sulfite solution (SO_3^{2-}(aq))

salt bridge

iodine solution (I_2(aq))

non-metal electrodes

The equations for the reactions taking place at the electrodes are:

$$SO_3^{2-}\ (aq) + H_2O\ (l) \longrightarrow SO_4^{2-}\ (aq) + 2H^+\ (aq) + 2e^-$$

$$I_2\ (l) + 2e^- \longrightarrow 2I^-\ (aq)$$

a Mark on the diagram the direction of electron flow.

1

b Combine the two equations to give the overall redox equation.

1

c What does the salt bridge allow to happen?

1

d State a non-metal which could be used for the electrodes.

1

39 Carbon has three isotopes. Carbon 12 and carbon 13 are stable.

The third isotope is carbon 14, which is unstable.

The decay of carbon 14 is shown below.

$$^{14}_{6}C \longrightarrow\ ^{14}_{7}N + X$$

a Name particle X.   1

b State **one** use of the carbon 14 isotope.   1

c Carbon 14 has a half life of 5730 years. How long would it take for a 20 g sample to decay to 2·5g?   1

d If a sample of carbon 14 was heated, what effect would this have on its half life? 1

40 Ammonia is an important starter material for the production of fertilisers.

The diagram below shows the stages in the production of ammonia.

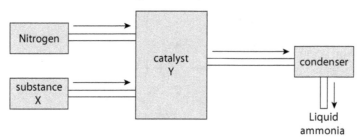

a Name the following: 1

Substance X   _____

Catalyst Y   _____

b Name the industrial process used to manufacture ammonia.   1

c The production of ammonia takes place at 400 to 450°C. Under these conditions
only 15% of the reactants convert to products.

   i A cooler temperature would see more reactants converting to products.
Why is a moderate rather than low temperature used?   1

   ii What step is taken by manufacturers to ensure all reactants eventually form
products?   1

d Ammonia is used in the manufacture of the fertiliser ammonium nitrate NH_4NO_3.
Calculate the percentage by mass of nitrogen in ammonium sulphate.   3

41 A student was given three unlabelled bottles.

Each bottle contained one of the oxides below.

      Potassium oxide   Copper oxide   Silver oxide

Using your knowledge of chemistry, describe how the pupil could determine which
bottle contained which oxide.   3

Graph paper

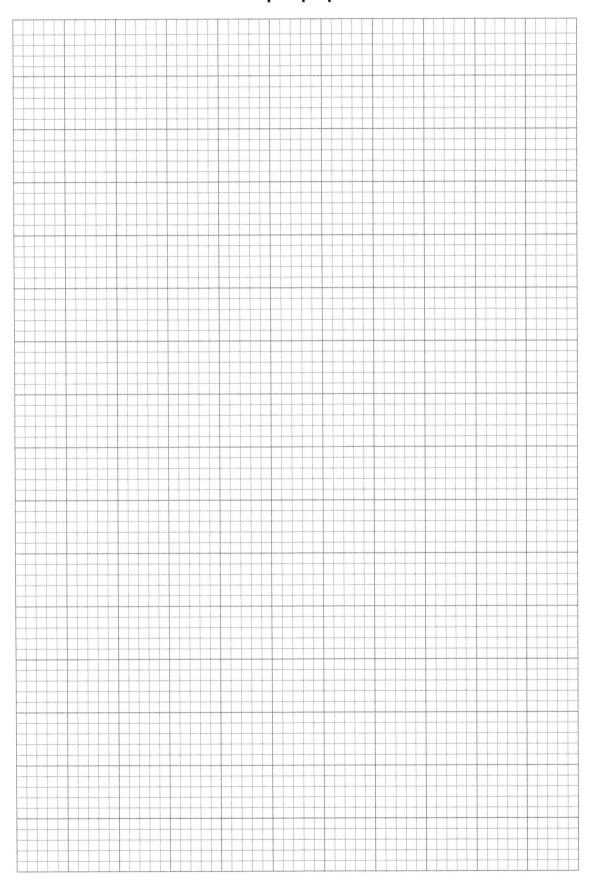

Graph paper

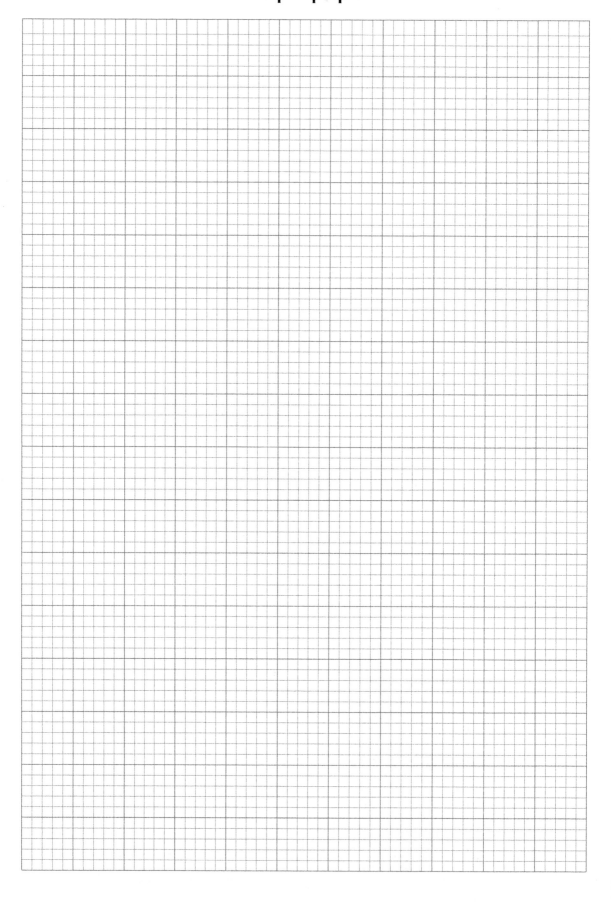